DARREN McGARVEY

The Social Distance Between Us

How Remote Politics Wrecked Britain

EBURY
PRESS

1

Ebury Press an imprint of Ebury Publishing,

20 Vauxhall Bridge Road,
London SW1V 2SA

Ebury Press is part of the Penguin Random House group of companies
whose addresses can be found at global.penguinrandomhouse.com

Penguin
Random House
UK

First published by Ebury Press in 2022
This edition published by Ebury Press in 2023

www.penguin.co.uk

A CIP catalogue record for this book is available from the British Library

ISBN 9781529103885

Typeset in 10.5/14 pt SabonLTStd by Jouve UK, Milton Keynes
Printed and bound in Great Britain by Clays Ltd, Elcograf S.p.A.

The authorised representative in the EEA is Penguin Random House Ireland,
Morrison Chambers, 32 Nassau Street, Dublin D02 YH68

In loving memory of Anna Wilson, beloved aunt, great-aunt, great-great-aunt and loyal friend to many.

'I do not think one can assess a writer's motives without knowing something of his early development. His subject matter will be determined by the age he lives in ... but before he ever begins to write he will have acquired an emotional attitude from which he will never completely escape.'

George Orwell, Why I Write, *1946*

Contents

CONTENTS

Introduction

'We didn't think there was going to be this lasting and permanent effect on sections of society. Nor did we foresee that
the benefits of the new economy were not going to be spread
out properly. The widening divide between those who really
thrived under the new economy, and those who suffered, the
gaps between the rich and the poor, which have got worse,
steadily, through Thatcher, Major, Blair and other governments over the last twenty odd years. We still haven't quite
worked out how to address that.'

*Kenneth Clarke (Chancellor of the Exchequer
from 1993 to 1997), Lord Speaker Lecture, 2021*

Britain is in a long-distance relationship with reality – Covid-19 only
brought this into sharper focus. As coronavirus ravaged the UK, stories of chaos, indecision and dysfunction at the very top of government
became commonplace. While it would be deeply naive to expect a
cumbersome bureaucracy such as the British government to respond
to a fast-moving crisis without making some mistakes, it was in the
UK that one of the world's worst pandemic responses was haphaz-
ardly mounted. One which provided a rare insight into how the
machinery of the state often operates behind the curtain of public
relations and how institutional intransigence and even the slightest
human error at the top of a democratic structure can adversely affect
millions of lives.

The prime minister himself initially regarded Covid as a 'scare
story' – a view quite typical of a man who has faced few impediments

to his progress in life. As Italy entered the first of many brutal lockdowns, its health systems overwhelmed as the death rate spiralled out of control, Boris Johnson remained characteristically cavalier in his attitude to the threat – allegedly even offering to be injected with the virus to prove it was harmless. Within weeks he was hospitalised.

As one of the Western world's most powerful leaders fell clumsily on his own sword, the profoundly grim nature of the crisis suddenly became tangible for many. A nation held its breath, in collective shock, as the initial novelty of life in lockdown, enabled by the illusory sense that the crisis was occurring somewhere else, wore off, replaced by uncertainty, anxiousness and fear. This was that rare occasion where, for once, a powerful political figure was no longer insulated from the impact of their decisions. No longer spared the consequences of their own poor judgement. Nobody was safe – not even the powerful. The virus's tentacles reached deep into the heart of government, humbling, humiliating and terrifying all who had doubted it before wresting them from their dream world where all that had mattered was what they believed.

While it's tempting – after years of grim pandemic-related news – to draw a line under everything that came before the pandemic and instead turn our eyes warily forward, there is a tremendous risk that in doing so we may leave ourselves even more vulnerable ahead of whichever crisis befalls us next. The 'normal' to which many of us would happily return in a heartbeat was anything but.

After a decade of austerity, Britain's public services lay decimated, 2 million children lived in poverty and food banks were the size of small factories. Thousands perished every year in drug-related deaths, tens of thousands were evicted from their homes and millions were either unemployed or trapped in the quicksand of in-work poverty. Media-generated public confusion and ignorance around issues like trauma, homelessness and addiction led to society's most vulnerable and persecuted groups being further marginalised while welfare reforms plunged thousands into destitution, driving spikes in rough-sleeping, mental health crises and suicides. Living costs increased as wages flatlined. Increasing numbers became trapped in a precarious labour market, designed to be flexible for exploitative employers. And as it emerged that the majority of children in poverty were being raised in households where at least one parent worked, the notion that honest graft and playing by the rules were enough to carve out a decent life in Britain was exposed as the pernicious lie it is.

Britain's social decline, by 2019, had become more visible in communities, as retail units lay empty on run-down high streets and waiting times for GP and hospital appointments increased. A chronic shortage of decent, affordable housing forced many into a cowboy rental market and deeper debt and as pressures grew on working people, so too did their anger and fear. Tabloids and daytime talk-show hosts happily supplied the demand for tougher action (and easy answers) by defaming the poorest communities as futureless and the most vulnerable and challenged families as scroungers and freeloaders. Economic insecurity and the sense of powerlessness felt across post-industrial Britain – rooted in longstanding social inequalities which preceded Britain's membership of the European Union by decades – led to a sharp rise in immigration concern, which was skilfully stoked by politicians and campaigners to deliver Brexit.

Meanwhile, a political class, which continued to enjoy generously subsidised travel, accommodation and food costs, handed itself consecutive pay rises as their axe of austerity continued to swing violently at public services. MPs are now paid over £16,000 more than they were in 2010, while nurses, teachers and doctors are, in effect, paid less. The poorest women and children have seen their benefits slashed as a shrinking group of multi-billionaires, enabled by lax systems of taxation, increased their lion's share of the spoils.

While it may comfort many to assume the sheer speed and scope of Covid-19, which swept the globe in a matter of weeks, lay at the root of the litany of governmental miscalculations and miscommunications that hampered Britain's initial response, in reality, the UK government's handling of the pandemic in those early weeks and months was not unusual in the slightest – if you understand how government really operates.

When Johnson's former advisor Dominic Cummings appeared before a select committee over a year after lockdowns became part of the British lexicon, one aspect of his scandalous, hyperbolic testimony rang worryingly true. Cummings's central allegation that officials in many senior positions within government and the civil service were clueless in the face of the crisis should not have come as a surprise to anyone. He described vividly the chaotic reality where chains of command and even the simple lay-outs of offices were conducive to little but the preservation of hierarchies and administrative status quos. He wasn't describing the British state in crisis – he was laying out how it actually works.

This farcical level of incompetence is not unique to the pandemic. The entitlement, the egos writing cheques that abilities couldn't cash, the scorn for expert advice, the jostling for personal and professional position and the tendency of many in high office to act in their own political interests and not in accordance with evidence are not pandemic-related anomalies – these are cornerstones of governance in Britain.

Leaders will make mistakes. That is part and parcel of living in a democracy. But we are entitled to expect that they learn from some of them. Britain's problems are fundamental and cannot be cynically chalked up to the pandemic.

The sense that our way of life is changing in alarming ways we do not fully comprehend has been palpable and widespread for years now. The pandemic, in many ways, was very on-brand for this phase in Britain's social, cultural and economic degeneration. Some believe the intertwining crises we are experiencing are economic in nature. Others that their roots are social and cultural. Some believe elites, dominating society from luxury cul-de-sacs, are responsible while others argue that globalisation and remote bureaucracies which govern from afar are the real culprits. Some allege that unwieldy and disruptive social media platforms, which have driven a firm wedge between people and the truth, bear responsibility for the curdling of public discourse, while prominent intellectuals claim there's never been a greater time to be alive. Powerful arguments are made from across the political spectrum that the true threat lies in a lurch towards populism while others warn of a vicious new strain of left-wing ideology, concocted on the cloistered, geographically remote campuses of elite universities, that threatens the very notion of objective truth. Maybe it's none of these things. Or perhaps it's a synthesis of all of them. Yet somehow, despite the dizzying abundance of information freely available, every piece of evidence set forth, rather than providing clarity, creates still more confusion, uncertainty and bitter disagreement.

Before the pandemic, Great Britain was already struggling to manage the myriad effects of rising poverty, employment insecurity, residential instability, low educational attainment and the attendant mental health problems, and drug- and alcohol-related death, dysfunction, and despair which accompany them. Covid-19 merely exacerbated the long-standing inequalities which have plagued Britain for decades. But whether you believe the issue lies with posh politicians who've never tasted desperation, ruthless exploitation by corporate elites or misguided, thin-skinned idealists, too short in the tooth to understand the real world, there is arguably one unifying theme which connects many of Britain's current difficulties – proximity.

Proximity is why you go to see your doctor in a surgery rather than chat over the phone. It's why journalists, police and lawyers like to speak to 'eye witnesses'. It's why our elected representatives congregate in the same location to debate and hold one another to account. Proximity is also why many people will trust the advice of someone who has encountered a problem they too are experiencing. Conversely, a lack of proximity is why many in the upper tiers of Britain's social order, whether in politics, media or business, are regarded with increasing levels of distrust by the wider population. It's why the Labour Party is increasingly loathed by the working-class communities it was founded to fight for. And a lack of proximity is why, in Britain, millions spoke of 'the end of the pandemic' when, in truth, the virus was only beginning to take hold in the poorer regions of the globe.

Proximity is how close we are to the action and how that affects the ways in which we assess, relate and respond to it. That might be a big-picture issue like poverty or a specific challenge like addiction – the closer we are to the conditions in which these issues tend to emerge, and the more familiar we are with the experiences of those who are challenged by them, the better placed we become to take decisions on how to address them. From wealthy politicians tackling poverty to drug counsellors who've never smoked a joint, and prominent commentators and celebrities opining on matters which do not sit within their experiential wheelhouse, Britain is riddled with problems of proximity across its key institutions and at every level of governance.

In complex societies, proximity has always presented challenges. Empires have fallen, rulers have been violently overthrown and countries have plunged into chaos in part due to the distance between the powerful and the powerless. Rather than the complex web of social policy, which can often be confusing, perhaps the clearest example of what happens when power lacks proximity is when geo-political foreign policy goes badly wrong when founded upon a lack of direct experience.

There are endless examples in recent history, from Vietnam to Iraq and Afghanistan. The presumption, as with powerful government's domestic policies, is always that a power with the world's richest resources and brightest brains at their disposal will surely do the best job possible. But often all these resources go to waste simply because local, lived experience and on-the-ground research were not sought out. Instead, these powerful forces are happier setting out a strategy based on their own strengths and biases. We look back now and balk at the fact the Iraqi army was disbanded by the United States at the start of the occupation in 2003. Almost 20 years later, we watched in disbelief as the Taliban retook Afghanistan within days of the withdrawal of Western forces.

Despite the trillions of dollars and endless manpower thrown into these countries, sustainable nation-building and workable futures can never be found because political and military leaders pursued confused strategies that totally failed to understand the motives of the local people and conditions on the ground – the diverse geography, languages, cultures, religions and history. Simple truths that anyone with a clear knowledge could have educated them on.

This example works too because in some senses, a distant government dominated by a political, managerial or class elite that tries to bring top-down, off-the-peg, ideologically driven policy into communities dominated by poverty is in many ways as much a colonial act as the worst examples of Western foreign policy. Indeed, when things go wrong with foreign policy it can resemble what goes wrong in misdirected domestic policy. Again, from Vietnam to Afghanistan, the American government poured in money to support proxy leadership that often wasn't representative of the majority of the population, which led to corruption and alienation.

Even worse, the process always allows the private sector to enter the fray, to both plunder resources and gain lucrative contracts managing the unstable situation they've created. And despite the futile repetition of this pattern, and the recurring theme of failure in its wake, successive leaders seem compelled, by institutional cultures which appear impervious to reform, to embark on the same disastrous courses of action as before.

Evidence exists that in every epoch, those regarded as knowledgeable and insightful, and thus granted the power to bring about change, often proceeded with courses of action which betrayed their ignorance, incompetence and self-interest – creating harm and misery for those they were trying to help. When those who possess the means to act upon society fail to comprehend fully enough the forces which comprise it, or the experiences of those who exist within it, they risk reproducing by orders of magnitude the very problems they are trying to solve.

Might Britain's many challenges in recent years, from the financial crash to austerity, have as much to do with the people we put in charge of tackling the problems as they do with the problems themselves? Ultimately, how do we reconcile the basic principle of representative democracy – where leaders are elected by popular vote – with the fact that most who ascend to high office are unrepresentative of most Britons?

We learned all about proximity during the pandemic when our distance from one another quickly became a matter of life and death. As Covid tore through the British population, resulting in one of the world's worst mortality rates, and the resulting economic destruction forced a Conservative government to throw out every fiscal rule in the book, the refrain that our leaders were 'following the science' was repeated daily to provide comfort – their way of reassuring us they were not, for once, following the politics. It was a rare admission that our democratically elected representatives understood very little. That, left to them, the country would quite simply have descended into mayhem.

Perhaps in the aftermath we should ponder the wisdom of leaving them to their own devices on so many of the other issues which affect our lives? Perhaps we should ask ourselves what exactly our leaders are qualified for, given the state the country was in before Covid even emerged in China. Could it be the case that before the onset of the pandemic, Britain found itself in such terrible social, economic and cultural shape in part because those same leaders, who would later turn in desperation to highly qualified experts in the face of a problem they could not comprehend, also lacked a deep enough understanding of the other pressing matters afflicting the country? Might the UK's vulnerability prior to the onset of coronavirus have had something to do with the dangerous lack of proximity so many in positions of authority have to longstanding and systemic problems like poverty? And might this lack of proximity, caused by relative privilege and limited social experience, extend also to those in politics and the media whom we trust to hold those in power to account? Could it be that Britain's problem is not that there is a lot of poverty but that we keep putting rich and powerful people in charge of sorting it out?

At what point do we look for the root of Britain's problems not in the lifestyle choices, attitudes and behaviours of working people, and the poor and the vulnerable, but in the groundless assumptions, false beliefs and prejudices of the apparently educated, cultured and sophisticated? Should we not also study those higher social castes, whose dominance of Britain's institutions has been virtually unbroken for centuries as we have stumbled from one crisis to the next? Might their biases, their hubris and their lack of humility play a small part in the UK's ceaseless psychodrama?

We elect politicians to advance our interests. But to advance the interests of any group, one must first understand something of those interests. How many MPs have experienced poverty? How many have had their toilet breaks timed, their in-work benefits slashed or the industries and amenities around which their communities were built closed down or demolished? How well-placed to address Britain's social housing crisis is a Parliament in which one in five MPs are landlords? How effective can a UK government, where 65 per cent of ministers were privately educated, ever be at closing the educational attainment gap? How much can a Chancellor of the Exchequer, estimated in the tabloids to be worth £200 million, married to a billionaire heiress, or a House of Commons where the vast majority of representatives come from middle-class backgrounds, ever truly understand either what poverty entails or how to solve it? And what exactly can a prime minister who spent £800 per week on takeaways during lockdown (while resisting campaigns to distribute free school meals in Britain's poorest areas) possibly grasp about the plight of malnourished, impoverished British children?

If the virus brought one thing into focus, it was surely that 'social distancing' – that is, the vast ravine of social experience between those who hold power and influence and those on the receiving end of their decisions – brought Britain to its knees long before the pandemic.

This is a book about distance. The distance between the powerful and powerless. The affluent and the poor. A book about how their interests and values diverge, and the assumptions they make about each other's experiences and intentions in the absence of any meaningful interaction from which to draw more accurate conclusions. The distance between media and the facts, and the chasms between the various political actors across the political spectrum and the working people they are currently making overtures to.

This book will apply the theme of proximity to a number of topics, and at various scales.

In Act One, we will examine the structural barriers in education, the labour market, criminal justice, property, immigration, housing, health and welfare which act together to entrench Britain's social class inequality – something many still claim doesn't exist. Blending the testimony of many courageous and insightful contributors, whose adversities I observed first-hand, with commentary, analysis and relevant aspects of my own lived experience, I will demonstrate not simply that class inequality exists but that it does so on a scale that only the wilfully ignorant could ever attempt to deny. I will show how social class inequality has cleaved the country in two and examine the cultural subtleties and sensitivities which arise in areas like language, media representation and attitudes to authority, as well as demonstrate how the privileged assumptions of the higher social castes, and the lopsided power dynamics from which they benefit, continually exacerbate and inflame the very ills too often blamed on the individual choices of the poor.

In Act Two, the book shifts from reportage and commentary to political and ideological analysis. We will explore conservativism, the radical Left, the roots of populism and the politics of the centre ground in an effort to understand how a lack of proximity influences policy design, campaigning and discourse. And we'll examine how social, economic or cultural remoteness means that even the noblest intentions can inadvertently cause harm.

Due to the constraints placed on my movements by lockdown, much of the content in this book takes place in Scotland, albeit with a few notable exceptions. Therefore, the chains of political accountability with respect to devolved issues (education, health, housing, etc.) lead rightly to the Scottish government. However, the problems we see north of the border are not unique to Scotland; they arise from UK-wide trends rooted firmly in Britain's economic and political structure. This is to say that administrations across the home nations are implicated to some degree or other in all of the class inequalities, structural or cultural, which lie at the root of the many issues outlined in this book. Those of a more politically tribal disposition should therefore resist the temptation to selectively ascribe blame and instead endeavour to reflect honestly – every administration plays politics with poverty.

This book was not written out of a desire to attribute Britain's problems to a singular source. It took over two years to write precisely because I was not interested in doing that. The truth is: there is plenty of blame to go around.

May this book serve as an earnest, occasionally pointed, attempt to bring some of you, kicking and screaming, if need be, a little closer to the action. And may it act as a call to arms for all of you who've been in the thick of it for too long.

ACT ONE

Proximity Warning

How systemic and cultural barriers maintain and expand Britain's historic social class inequalities

I

Not in the Same Class
Bootcamp for class denialists

At a certain age, being taken into police custody is a rite of passage where I'm from. There came times in my life when I probably deserved being thrown in the cells but the very first time it happened, in 2012, I certainly did not. I had been partying for two days in a flat in Garnethill, Glasgow. This is a part of town not often associated with an overbearing police presence, which is perhaps why so many middle-class people felt comfortable taking or dealing drugs there. With its immaculate tenements, private school and stunning views of the north of the city, Garnethill attracts students with reasonably well-off parents and young, affluent professionals. Naturally, when I started seeing a musician who rented a spacious corner flat situated on the unofficial border between this highly coveted quarter and the rather less plush, chewing-gum gauntlet of the city centre, I assumed, naively, that I was once again moving up in the world.

It was almost time to throw in the towel after another successful two-day bender. We were about to fall into bed when I realised that I had run out of cigarettes and would not make it through the night terrors without some. Fortunately, there was no shortage of late-night shops to acquire them. Though when I arrived at the local down the road I found it closed, meaning I had to double-back on myself and head into the maelstrom of Glasgow's infamously rowdy nightlife. As I turned around, I noticed a police van on the corner.

For those of a certain age and class, the sight of law enforcement produces anxiety which can affect how you behave. You may begin to appear worthier of suspicion than you otherwise would have, had you not just seen a meat-wagon. As a result of a nervier demeanour, you become likelier to draw the police's unwelcome and steely gaze. I tried to play it cool but, alas, the cops exited the vehicle and walked towards me, looking all friendly at first, as they often do. Then, the standard questions: how are you? Where are you going? What's your name? I made the classic error of cooperating. They searched me, then ran my name through the system and, to my dismay, placed me immediately under arrest.

We arrived at the station and I was escorted into the building to be processed, before being placed in a cell – on Saturday night. 'Fucking scummy bastards,' screamed a woman along the hall, punching the door in a blood-curdling rage. Officers laughed it off but that seemed to antagonise her further. The station was busy. There was a sinister buzz about the place and the very palpable sense that if not for the massive steel doors, and sophisticated tools of restraint at the officers' disposal, they would be quickly overrun.

Given this was my first night in a cell, I made the mistake of assuming whatever mix-up had occurred on the part of the police would be quickly detected and rectified. I, after all, was a respectable citizen who had never run into trouble with the law. One hour passed. Then two. I searched my pockets hoping to locate some chemical particle. Nothing. Three hours passed. Then four. The effects of the alcohol had by then worn off, leaving me restless and agitated.

In spite of the noise, I drifted off. I was awakened hours later, disorientated, by a startling knock of the door. There was no way to ascertain the time of day. A plate of sausage and beans was tossed into the cell, which I promptly devoured. Eventually, the cell door was opened and I was handed a phone. It was a lawyer. The angry-sounding man appeared to assume I was either stupid or illiterate. He explained, in a tone which perfectly braided aggression and condescension, that there had been a warrant out for my arrest for some time and that my failure to present at a police station was why I had been detained. I explained that I was not aware such a warrant existed. Had I been, I would have been more than happy to cooperate. It turned out the letter had been sent to a previous address. I was advised by the lawyer to respond only with the words 'no comment' when questioned and that under no circumstances should I say anything else. I said goodbye and a cop returned, took the phone and slammed the cell door shut behind him. Within an hour, I was taken to a room for questioning.

Two officers sat opposite me. Normal guys, watching the clock. The previous 12 hours had been one of the most frustrating episodes of my life but for them it had been just another long and boring shift. Their veneer of power faded – I was sober and understood perfectly what was going on. This was a game.

'Are you Darren McGarvey?'

'No comment.'

'Do you own a green jacket with a hood?'

'No comment.'

The questions continued, each one as absurd as the last. 'Do you upload YouTube videos under the name Loki?'

At this point, I could barely contain a smirk. 'No comment.'

It wasn't the questions which were laughable but the dance we were doing. I was detained because an administrative error on the part of the police had led to a warrant being posted to the wrong address, yet I was being probed like my failure to present at a police station was either my fault or an admission of guilt.

'No comment.' 'No comment.' 'No comment.'

Theft by house break was the crime I was suspected of. A crime I had never heard of let alone committed – unless we're counting the numerous occasions I drunkenly broke into my own flat, having lost my key. I was, nonetheless, the last person to be released from the station – within seconds of the 12-hour limit at which a person can be legally detained without charge. Bastards.

Had I left the flat the previous night in pretentious spectacles, a three-piece suit and loafers, would I have been stopped? I very much doubt it. Unfortunately for me, I was wearing my 'housing estate uniform' – a pair of jogging bottoms, trainers and a hoody – like everyone else who was released before me. Had I invoked my legal right not to be searched and denied the request to give my name, might I have been in bed, with a 20 deck of cigarettes, by 11.30pm? My central error, in this instance, was my investment in the ludicrous notion that cooperating with law enforcement is a good idea when you have not broken the law – a most serious misapprehension.

It is perfectly fair that an individual, reasonably suspected of criminality, might be stopped, searched or even detained by police officers. I was, however, stopped at complete random because I fitted a profile. They weren't out looking for me; I simply looked like the kind of person the police are always trying to catch. My lack of experience in dealing with cops until then was why I didn't tell them to run a mile when they asked for my name. How many young people in the UK, by virtue of their appearance, draw the attention of law enforcement, like I did, irrespective of the illegality of their behaviour, and are then submitted to similar processes (and errors) as I was? And how many, as a result of their mistreatment by police, become pathologically suspicious of law enforcement?

Police have two faces – one for the people they serve and one for the people they pursue. I'm rare in that I've had a good look at both and could confidently pick them out of a line-up. At some point, because I was regularly swanning around plusher parts of Glasgow, like the West End and Garnethill, riding on the coattails of middle-class friends and lovers, I had begun to wrongly assume that I was now cut from a finer cloth than my plebian friends of old. But when push came to shove, in the eyes of the law, I was just another pleb. Slim pickings, there for the taking. Nothing arrests your assumptions about class like that first long night in the cells.

Three weeks into a self-imposed news and social media ban in 2019, with a criminal conviction finally under my belt, I was drawn, perhaps inevitably, back through the dreaded looking glass by two simple words – class war. They boomed from a radio speaker and, like the cries of my children, were impossible to ignore.

It is highly unusual to hear the topic of class being discussed in the mainstream news in anything but the vaguest terms. Even rarer to hear it discussed in the context of war. It was rather exciting, truth be told. ITV's Robert Peston – a man so middle class he ought to have a reed diffuser scent named after him – seemed somewhat alarmed by the Labour Party's 2019 manifesto. This was a time before Covid, when free lunches provided by the state were akin to authoritarianism. Labour's pledge to confront the so-called digital divide – the growing disparity in access to technology and thus information, education and opportunity between social classes – by providing every household in the UK with free internet therefore set alarm bells ringing. It was one of many policies which aimed to redistribute opportunity more evenly between the affluent and the less well-off which, on its own, wasn't that radical but when lumped together with a bunch of other distinctly left-wing proposals, in one manifesto, was never going to make it past the London media establishment. Like many journalists working under extreme time pressure, and understandably keen to frame the story, Peston slipped into sensationalism. Writing in the *Spectator*, he declared that Labour leader Jeremy Corbyn, by publishing some policies which aimed to tackle inequality on some pieces of paper, was waging a 'class war'. In a sense he was right, the only thing he failed to point out was that Corbyn's Labour Party was attempting to launch a retaliatory strike – the Conservative Party's class war of austerity had raged for ten long years prior. This typically hot take was tumbling from the mouth of the man who, perhaps more than any other British journalist, popularised the term 'credit crunch' – an Orwellian political phrase designed to obscure how the deregulated financial services sector wrecked the global economy by overselling poor people houses they couldn't afford. Peston, perhaps shaken by the communist document, seemed suddenly struck by a nightmarish vision of Britain in Venezuelan freefall, where the internet was a public service, Jeff Bezos paid a higher marginal tax rate than his personal assistants and sexual assault survivors didn't have to prove they were raped to get child benefit.

This insistence that Labour's policy proposals were a declaration of class warfare was, to my mind, extremely odd; few journalists had mentioned class war at all since the economic crash, after which the newly elected coalition government ushered in the deepest public sector cuts of any advanced economy. Nor could I recall so much as whisper in the language of class warfare when food-bank use, homelessness and suicides appeared to increase sharply around the same time the harshest welfare reforms in British history had begun to take effect.

It wasn't 'class war' when the publication of the Panama Papers revealed a network of banks, monarchs, heads of state and organised criminals engaged in the plundering of trillions offshore. It wasn't 'class war' when the UN revealed that the richest 26 people in the world held more wealth than half of the global population. And it wasn't called 'class war' when the captains of high finance, drunk at the wheel before crashing the economy, were bailed out before getting barely a slap on the wrist – while the poorest received arbitrary sanctions for arriving a few minutes late to appointments at the job centre. Not even the Grenfell Tower fire – a direct consequence of the class inequality between wealthy, politically connected property owners and working class, politically excluded ethnic minorities – was enough to inspire Peston into the sort of irrational, Oxbridge prefect hysteria elicited by Corbyn's 'free internet' bombshell. It was, however, 'class war' when the interests of the affluent and wealthy were temporarily dislodged from their near permanent place atop the hierarchy of mainstream political concerns.

Peston has his talents (as do the great majority of journalists working today) but his class analysis was incoherent. This, of course, should come as no surprise. Ignorance around the issue of social class is, like opportunity and wealth, highly concentrated near the top. The Sutton Trust, a prominent social mobility think-tank, found in 2016 that 51 per cent of the country's leading journalists were educated privately and just under 80 per cent of its top editors went to either private or grammar schools. Expecting the beneficiaries of the inequity which characterises the British class system to be capable of thinking or reporting objectively about it is a bit like expecting a benzodiazepine addict to write their own Valium prescriptions. And when the parameters of any discussion or investigation by media on the topic of class are so often set by those who fail to perceive it, due to their relative distance from its negative effects, it is no surprise that, whenever the issue does occasionally surface, it is almost always framed in narrow or superficial terms that render any subsequent discourse pointless – or infuriating.

In recent years, the language of class has been all but wiped from our lexicon. In its place, terms like 'social mobility', 'poverty of aspiration' and 'meritocracy' have taken centre stage – each an attempt to explain the vast differentials in outcomes between individuals and communities based on their social backgrounds without invoking the coarse and apparently divisive language of class. This shrouding of basic truths where social class is concerned, while not always deliberate, has potentially serious implications. Not least when it extends to the economy – the mechanism by which inequality is either reduced or widened – which is increasingly referred to in figurative terms that fail to convey its true nature.

'Six years on from the financial crash that brought the world to its knees,' wrote then-Prime Minister David Cameron in the *Guardian* in 2014, 'red warning lights are once again flashing on the dashboard of the global economy.' The stark warning came following a meeting of the G20 in Brisbane as the Eurozone lay on the brink of recession. Cameron's description of the global economy as akin to the cockpit of an aircraft was effective at conveying the gravity of the situation. Given the dire state of the economic affairs in the previous six years since the financial crash in 2008, Cameron did well to conjure a semi-original figure of speech; the best apocalyptic metaphors had by then already been deployed by his chancellor, George Osborne. Speaking at the 2011 Conservative Party conference, Osborne promised we would 'ride out the storm' – which heavily implied that the economy was an unwieldy force of nature, beyond the influence of mere men, which, rather than confronted, could only be endured.

Over the years, economies have been depicted by leaders as broken-down vehicles, patients on sick beds, wars, sports, gardens, pies and purses. A great figure of speech is undoubtedly useful in illustrating basic economic principles and signalling the general health of a nation's finances but too often, whichever framing device a politician uses, it invariably fails to capture the economic reality in Britain.

You don't have to be as far left as Jeremy Corbyn to recognise how fundamentally unfair the British economic system is, yet leaders are consistently keen to portray economic turbulence as phenomena that impact everyone equally. That we are all part of the same team, swimming in the same direction, when this is simply untrue.

In a recent interview, US tycoon and billionaire investor Charlie Munger shared his thoughts on navigating the unpredictable landscape of the free market. His comments revealed the stark contrast in how economic fluctuations are perceived and experienced depending on your socioeconomic status. 'You know, the economy sometimes booms and sometimes it doesn't,' he told Yahoo News in May 2019, 'and you have to live your life through both episodes. And our attitude is, we just keep swimming. Sometimes the tide is with us and sometimes against, but we keep swimming either way.'

Munger was, of course, describing the boom-and-bust economic cycle as it is experienced by the billionaire, asset-owning class. From his vantage point, the maritime metaphors make sense. During times of recession, figures like Munger possess the resources, knowledge and experience to not simply remain afloat but to gracefully ride the crest of every wave. From his socioeconomic position, recessions separate the men from the boys. They are inevitable, naturally occurring events which, rather than avoided, must be embraced. But for people who do not own anything, whose economic survival depends on competing for increasingly insecure, precarious and low-paid work, the economy is not so much a choppy sea to be skilfully navigated but a pool of turgid quicksand in which remaining completely still and taking as few risks as possible is the only way to survive. When a recession begins to bite, these people lose their jobs. They default on their debts. Their families are placed under intolerable mental and psychological strain. Some may end up out of work for so long that they slip into mental health problems, addictions, welfare dependency and even criminality. If there are children in the mix when this occurs, they too become likelier to experience the adversities endured by their parents, such is the nature of generational poverty.

And at every intersection between the disadvantaged and the market, an enterprise stands ready to extract profit or opportunity from another person's sharp descent. The pharmaceutical companies, the alcohol manufacturers, the pay-day-loan sharks, the junk-food producers and the ubiquitous bookmakers – when calamity strikes it become a licence to print money. The Mungers of the world derive economic resilience from the diversity of their portfolios, which leaves them relatively unexposed to downturns, while entire working-class communities are dangerously dependent on the presence of one or two precarious industries to wrest themselves up and out of the bog.

It is, of course, perfectly understandable that a man of Munger's stature would speak in these terms: from his vantage point, this conception of what economic turbulence looks and feels like is a matter of intuition, rooted in his experience. The problems arise when figures like Munger, as a result of their wealth, grow increasingly remote from the rest of the population. When their interests, aspirations and values diverge so dramatically from those further down the food chain that they, in many ways, become antagonistic. When some win big in the casino of the global economy, others lose everything. And in Britain, when political leaders like Cameron frame economic issues for journalists like Peston in the floweriest possible terms, the information eventually relayed to the general public is as good as useless – if the ultimate aim is to arm people with facts about the material conditions of their lives and not patronising, bitesize narratives. What an economic metaphor deployed by a politician almost always fails to convey is the fact that we are not all on same side.

Can you imagine if a world leader went on television during a crisis and said something along the lines of the following: 'The car park of the economy has very limited spaces because that is what works best for the owners, so during the next few years of mayhem, you lot down there will just have to buckle up and hope for the best. If you can't locate a space, that's your fault – try harder. And can you keep the noise down? We're counting a lot of money up here.'

You might think that a cynical analogy but such a view may be rooted in your favourable experience of the economy. One which has led you to conclude that things are not that bad. That the economy we have represents the apex of what is economically possible. Of course, a toxic by-product of structural inequality is the great disagreement between different sections of society about the nature of Britain's economic system. Disagreement which arises from parallel but often opposite experiences. Some regard the UK economy as a high-water mark in the development of civilisation while others believe it's limping in its agonal stages before assuming it's final zombie form. The point is: just because the economy is not bad for you doesn't necessarily mean it's good for someone else. Public discourse around economics tends to be framed by those for whom economic decisions are not matters of life and death – this is a problem.

Fundamentally, what is beneficial for the owner of the car park is not always beneficial for the drivers who use it. These competing interests are, loosely speaking, the essence of class conflict and are replicated across the economic landscape in the labour and housing markets, in policing and criminal justice, and, more subtly, in educational and health systems. These competing interests – and the fact they cannot always be reconciled – rarely change. Politics exists primarily to square them off but increasingly, the political class appears more keenly in tune with the interests of the affluent because those are interests which, in many ways, they share.

What politicians fail to convey when they speak in figurative terms about the economy (and what journalists tasked with scrutinising them must consider more deeply going forward) is that class inequality in Britain is not some freakish event – it's how our economic system justifies itself in the eyes of those whose interests it is primarily configured to serve. Private profit often comes at the expense of public health and wellbeing. Private enterprise often expands at the expense of a shrinking public realm. When economic growth so often hinges on constraining the wages, standards of living or opportunities of working-class people, conflict becomes inevitable. Most people are OK with a certain degree of inequality and few would argue that equality of outcome across the board is achievable or even desirable, but the galling and unsustainable disparity between social classes which has occurred in the last 40 or so years in Britain, on the watches of highly educated politicians and under the noses of a great many intrepid journalistic juggernauts, goes far beyond what is reasonable, natural or fair. Our system is now, in many ways, gamed to protect the winners while over-penalising working people, the poor and the vulnerable. As a result, society is now set in two distinct and parallel economic realities, from which entirely divergent interests and cultures have emerged, and between which conflict has become politically unmanageable.

When Cameron made his pronouncement in 2014, the harsh austerity measures his government had brought forward in order to balance the books had begun to take effect. Food poverty, homelessness, drug deaths and suicides were on the rise. Despite the financial crisis being caused, primarily, by wealthy bankers and speculators who conflated their own distinct economic interests with those of wider society, and who were enabled and encouraged by political leaders who deregulated the economy under their self-serving advice, the poorest in society bore the brunt of paying for the crisis in the form of unemployment, public sector cuts and an almost permanent sense of economic insecurity. And while Cameron warned the country to brace for another impact in 2014, a massive transfer of wealth upwards was already well under way – which both he and Peston said surprisingly little about.

According to analysis published by the Equality Trust, Britain's billionaires would, within a year of Cameron's *Guardian* article, see their net worth more than double since the financial crash, with the richest 1,000 families amassing for themselves a total of £547 billion – a controlling stake in the business of society. By 2015 – a year after Cameron's article – the Equality Trust revealed that these richest 1,000 families had accumulated as much wealth as 40 per cent of the UK population, seeing their wealth increase by £28 billion – £77 million a day. Did some rich people lose money in the crash? Yes, of course. Did some poor people find a way up the ladder? Yes, of course. But neither of those facts negates the broader truth – the manner in which the economy is constantly framed by political leaders and media pundits bears little relation to the reality experienced by the millions who have not shared in the prosperity. As of 2021, the combined riches of the world's ten wealthiest men rose by £400 billion during the pandemic, according to a press release issued by Oxfam.

The charity claimed this amount would have been enough to prevent the world from descending into poverty because of the virus and sufficient to provide vaccines for every person on the planet. While poverty is obviously a little more complex than mere wealth redistribution, it is clear the vulgar wealth hoarding (and the political imbalance it creates) not only defines the current economic period but is a key aggravating factor in social inequality generally.

Britain's central problem is class inequality and the distance it drives between those who have done well under the current economic settlement and those who are suffering because of it. What better illustrates the divergent interests of the powerful and the powerless than the sorry sight of David Cameron being hauled out in front of a select committee in 2021 to explain why he used his position as a former prime minister to attempt to advance the interests of his wealthy friends and associates, on at least 47 occasions, during a pandemic which disproportionately affected the poor? Cameron was, in his mind, simply following his economic interests and those of his friends, in accordance with the law. Perhaps he even believed his actions would benefit those struggling, in some abstract way. But for many, looking on from a distance, his behaviour appeared almost as crooked as the rules themselves.

*

What do we mean by class? Is it defined by how we feel about ourselves or our backgrounds? Can our social class be gleaned from how we speak, where we live or the newspapers we read? Or is there a more objective determining factor which places us squarely within a certain socioeconomic grouping? The *Encyclopaedia Britannica* defines social class as 'a group of people within a society who possess the same socioeconomic status'. In sociology, people in the same social class are typically said to share a similar level of wealth, educational attainment, job type and income. In media, news or advertising, as well as in market research, we find perhaps the most widely used method of social classification in the UK is the National Readership Survey's social grades system, the distinguishing feature of which is that it is based solely on occupation.

When we think about our jobs, we think about what we do, physically and mentally, and what we are paid in return. But the means by which we make a living represents just one facet of a far more complex relationship with the economy. Who do we work for? What goods and services does our work ultimately produce? What value is placed on our contribution to this enterprise, who or what determines this value and is that value a fair reflection of either how hard we work or the true worth of that labour?

Think of the key workers we applauded during the pandemic. The misty-eyed pronouncements of political leaders hailing them as the backbone of our entire way of life during the crisis. And now think of how many of them either lost their jobs, had to change occupation or were forced to threaten (or resort to) industrial action in the face of low pay and poor working conditions – as the billionaire class entered a new space race.

People of all social classes face adversities; wealth cannot insulate us from relationship breakdowns, illness, mental health problems or bereavements. But a person's social class is defined, essentially, by how much they have to struggle day-to-day simply to meet their basic needs – as well as those inevitable adversities. For some, this burden is simply immense and quite frankly unjust when considering the levels of wealth their collective labour generates for others or, in the case of key workers like nurses or bus drivers, how integral they are to ensuring our way of life is maintained.

Some of you don't want to talk about class. Some of you think it irrelevant. Worse still, among you, there are those who claim class does not exist. When people say they don't believe in class, we must conclude they've arrived at such a position either because they do not see it or don't want to see it. While some forms of denial are more forgivable than others, our failure as a society to acknowledge the role class plays, not simply in determining key outcomes for individuals and communities but in setting the tone and trajectory of our society, is not only wrong but dangerous – for all social classes. The tremendous irony is that those who would prefer the topic was no longer discussed, or believe it does not exist, often harbour the very attitudes, false beliefs and prejudices, and are engaged in the very economic practices, that virtually assure the debate about class will persist. Those who insisted not so long ago that the class war was over may be in for a big surprise.

The concept of class has become synonymous with the far left and, for some, a by-word for Marxism, the implication being that communist thinkers created it in a deliberate attempt to stoke social division, ripening the conditions for a violent revolution. However, this argument that class was 'invented' by Marx and Engels in the mid-nineteenth century, does not deal with the economic and social forces which preceded Marxist thought or the philosophical inquiries which shaped and influenced it. It ignores the conflicts of interest between barons and the Crown which led to the drafting of the Magna Carta, which limited the power of monarchy. And it does not consider the socioeconomic forces which led to the Peasants' Revolt of 1381, when agricultural labourers and the urban working classes united to resist an unpopular poll tax, executing merchants, and razing a royal palace, to the horror of a young Richard II.

From the aftermath of the English Civil War to the onset of the Industrial Revolution, class was a central factor in how wealth and opportunity were distributed. It was also central to landmark reforms such as the abolition of slavery, the right to vote and the creation of the Welfare State. Claiming class was breathed into existence by philosophical word-of-mouth rather than instituted through centuries of political, social and economic development and struggle is rather like insisting gravity didn't exist until a famous physicist gave it a name. And this tedious and juvenile notion that class is not relevant because the average person is not explicitly preoccupied by it on a daily basis is just as nonsensical – people in Hartlepool aren't talking about Isaac Newton but this has no bearing on the tremendous force classical mechanics exerts over their lives.

We can debate how class should and should not be applied. We may even argue in good faith about the appropriate tone such a debate ought to have, but the discourse around whether or not it exists is a diversion unworthy of any serious person's time. The concept of class arose and has persisted despite concerted efforts to flush it from public consciousness not because it creates social division but because it so accurately describes the forces which drive it. Accepting that is not radical in the slightest – it's a commitment to grounding the discussion about inequality well within the bounds of social reality.

In the next few chapters, we will focus on some of the institutional class barriers which exist across British society in an attempt to nest a potentially abstract discussion in something tangible. We will examine various structural obstacles which those from lower social classes often come up against, that constrain or impede entirely their social mobility, and how those from higher social classes who face fewer, or indeed no such barriers, are often unconsciously complicit in their persistence.

But let us first establish briefly one solid example of a structural class barrier. Like many barriers, this one is so pervasive yet so elusive that many of you, irrespective of your background, may not yet have even identified it – appeals processes. In 2019, around half of people with disabilities who appealed a Department of Work and Pensions (DWP) decision were successful. The BBC reported that, in total, more than 550,000 people won an appeal over their benefits at tribunal between 2013 and 2018. Many successfully appealed thanks to welfare rights advisors, who supported claimants in navigating a deliberately opaque and hostile appeals process. We will explore welfare in greater depth later, but for now let's consider why half a million disabled people were wrongly denied benefits it subsequently transpired they were legally entitled to? Could wrongly denying vulnerable, disabled people their benefits, rather than administrative oversight, in fact be an institutional strategy? One rooted in the understanding that many in this disadvantaged group have not the means, time, nor headspace to take the immense risk of challenging the DWP? Whether applying for benefits, being given community service or being abruptly sacked by an employer, citizens are often assured of their right to 'appeal'. Sometimes, they may even be sent a nice letter encouraging them to do so. But despite this legal right, many do not pursue one – this is well understood and it's partly on that basis that they are offered one.

Appeals processes, while giving the outward appearance of being inclusive and democratic, often act as filtering mechanisms which deter disadvantaged people in areas like criminal justice, welfare, employment and education. They are often arduous, time-consuming, stressful and costly endeavours. It therefore follows that those who possess the resources to fully engage with such processes stand a greater chance of successfully appealing than those who do not – this is an example of a structural class barrier.

Across British society, similarly systemic quirks can be found, which flush poorer people out, creating more room for the sharp elbows of the affluent. Appeals processes are just one example in an exhaustive list of structural class barriers across health, housing, criminal justice, property and education, which arise when a system is configured, administered and facilitated by and for one social class. A system which intuitively serves the interests of wealthier citizens while simultaneously obstructing working-class people and pushing the poorest further down, while failing to perceive that this injustice is occurring. The lowlier your class position in society, the more barriers you face and the harder time you have getting anyone in a position of influence to do anything about it. In isolation, some structural barriers may not seem insurmountable, but taken together, they make life difficult for people from poorer backgrounds and intersect to create the conditions for Britain's unsustainable institutional class inequality.

The billionaire investor and philanthropist Warren Buffett once warned the *New York Times*, 'There's class warfare, alright, it's my class, the rich class, that's making war and we're winning.' He was and remains correct. And yet, few can truly comprehend just how little effort is truly required on the part of the well-off to dominate the poor (being that the systemic scales are tipped forever in the favour of the prosperous) or, indeed, how much toil this imbalance thrusts upon the shoulders of workers, the vulnerable and the poor.

2

Second Class Citizens
The employment and education gulfs

> 'We want one class of persons to have a liberal education, and we
> want another class of persons, a very much larger class, of neces-
> sity, in every society, to forgo the privileges of a liberal education
> and fit themselves to perform specific difficult manual tasks.'
>
> *Woodrow Wilson (then-president of Princeton*
> *University), address to the New York City*
> *High School Teachers Association, 1909*

Every day's a school day until you have to get a job. My experience of
employment started when I was 17 years old, after leaving education
abruptly following a bereavement. I began working as a kitchen porter
for an agency. I didn't have a minimum set of hours and therefore my
income fluctuated. Thankfully, living with my grandparents at the time,
I wasn't exposed to the potential stress such an arrangement might
cause – I was working for extra cash and not to meet my basic needs.

Sometimes I'd get a placement that lasted a few weeks, other times
I'd get a call in the morning to tell me to be at a certain place by a certain time. I detested the latter because as well as the tough nature of
the work itself (washing dishes and large manky pots, mopping and
scrubbing greasy floors and surfaces), every shift was like going in nervously on your first day. You'd have to meet a new boss, new work-
mates and learn what the job entailed in a particular workplace, over
and over again. I worked in hospitals, schools and even in the kitchens
of some of the broadcasters I'd later write and present documentaries
for. I enjoyed the fact that I was working (it got my grandad off my

back) but the work itself was difficult to relish as the temporary nature of the gigs kept you trapped on low pay with little possibility of progression. And, due to being at the impromptu behest of the agency, I had no room to plan any further than the next few days.

Lucky, then, that I was just a teenager. And while I had left school due to the impact of my mother's sudden passing and was dealing with mental health issues arising from the grief, I had no real financial responsibility. Had I been a young father however, like my dad when I was born, then this kind of work, straight out of school, while trying to process pain and trauma, estranged from family, would have been tremendously difficult to build any sort of life around.

Our social class is defined, more than any other aspect of our lives, by what we do for a living. It's from certain types of skilled and unskilled occupations that the other factors with which class is often associated (earnings, lifestyle, consumer habits, whether we own property and where we holiday) begin to open up to us. Naturally, wealth and prosperity, or working in a prestigious profession, come with a level of social respectability. One which, in my experience, is not always conferred on taxi drivers, checkout assistants, refuse workers or kitchen porters.

In the not-so-distant past, being working class was a source of great pride and dignity for many. This is not to say that is not still the case now but today there seems to be immense confusion about what class means; the aim of the game seems to be to rise up and out of your class, not simply for the kudos of being seen as a 'success' but because remaining working class these days is, in many ways, resigning yourself to certain hardships – job insecurity, low pay, few paths to career progression, the likelihood of your offspring becoming consigned to the same fate and a general sense of being looked down upon. In 1999, when Tony Blair proclaimed, 'I want to make you all middle class,' his remark was regarded as forward looking and ambitious – but examined from another angle, you might say it revealed how he perceived working classness as a phase to be transitioned out of. All the more bizarre given his insistence that class was a thing of the past.

Before the onset of deindustrialisation, a job which involved the performance of 'difficult manual tasks', such as one in the shipbuilding industry, was tough but could pay enough to feed a family and run a household. As well as paying enough to live on, the goods produced were physical and served a tangible purpose in wider society, which meant the tough labour brought with it a sense of dignity, pride, and purpose. While these jobs were often arduous and dangerous, and the hours long, collective bargaining offered a level of protection against exploitation and unsafe conditions and the booming industry meant many employers competed fiercely to attract and retain workers. We may look back at those times through rose-tinted glasses but there's an obvious reason for that: a job was virtually guaranteed and enough to raise a family and sustain you throughout your working life.

Today, things operate a little differently. A full-time job is no longer a guarantee you'll have enough to live on, nor are two or three part-time ones. Most people who claim benefits are in work, meaning employment is no longer a route out of poverty and that the Welfare State actively subsidises and, indeed, incentivises low wages. Many of these low-paid jobs are in the service sector and involve zero-hours contracts and intense, rotational shift patterns which make planning ahead difficult. Workers don't always derive the same sense of purpose and dignity from plying their trades as the output of their labour is often less tangible and much of it is performed in various forms of isolation.

The most significant shift in the labour market in recent times with respect to working-class occupations, and obviously apart from technological and digital advances, was the marginalisation of trade unions, beginning in the 1980s, which has since contributed not simply to the ascent of precarious work but also a depressing resignation, acceptance even, among vast swathes of a young twenty-first-century workforce that insecure work and poverty wages are normal. No industry embodies this decline more than the call centre sector.

Pre-pandemic, around 850,000 workers were employed across 6,000 call centres in the UK (according to a report 'Well-being and Call Centres' from the Institute for Employment Studies). During the pandemic, the *MailOnline*'s money section reported that this number surged to around 1.3 million – 4 per cent of the entire UK workforce. The majority of call-handlers in the sector are working class and this status is implied not simply by what they earn – the average salary is £17,970 a year – but also by what is expected of them by their employers and how they are treated when they go to work.

'Emotionally, it's a struggle,' a middle-aged, female call-handler from

Greenock told me in spring 2021, when I visited a call centre hotspot to learn more about the standard practices in the sector. Like many former shipbuilding strongholds, Inverclyde, Greenock's local authority area, has become attractive to companies due to high levels of unemployment and the incentives offered by councils in return for investment.

'It's certainly not a job for someone that's faint-hearted, or not got thick skin' she explained. For her protection, I will not reveal her identity – she fears repercussions for sharing her experience of working in call centres for over 20 years. 'There's barely a week that goes by when you're not sworn at or vilified for your accent,' she said, with an oddly detached frankness, suggesting she has grown so accustomed to being verbally abused that it no longer shocks her. It's this sense of resignation in the face of such poor working conditions that this sector (and countless others which depend on flexible working-class labour) seeks to cultivate through its workplace practices.

The pressure of the call centre, whether telemarketing or providing a point-of-contact for public or private services, goes beyond the calls themselves, though they certainly are a challenge at times. The additional strain often comes from how relentlessly programmed and tightly monitored the average working day is, to ensure maximum productivity. 'Every minute of your day is scheduled,' she explained. A worker's scheduled tasks, breaks and even toilet trips are tracked, logged and analysed. 'It's common to be told you owe back a minute just because you were logged off for too long,' she told me. What happens if she's two minutes late to her workstation following a queue for the toilet? 'So, I'll need to put two minutes extra at the end of my shift and make it up – the minute I am not where I am supposed to be, I'll be flashing up on somebody's screen in red.'

We were joined by Craig Anderson, the regional secretary for Communication Workers in Scotland, the union that represents call centre staff, and Phil Taylor, professor of work and employment studies at Strathclyde Business School, who's been studying the call centre sector for as long as our friend's been employed by it. 'The cost-minimisation imperative that drives call centres means that you get large numbers of people subject to really hard performance targets that measure average call time, wrap time, toilet breaks, the lot, down to hundredths of a second,' Phil explained, his visible passion unconstrained by the academic language. The level of surveillance that exists in many call centres is extreme, and akin to having someone standing over your shoulder throughout the course of each shift, which can be as long as 12 hours, over many consecutive days.

You might think talking on the phone can't be that hard but you'd be wrong. Call centre workers have an intense level of social contact with their customers. They often deal with angry or unpleasant people and so their jobs involve a baseline level of conflict which requires psychological and emotional resilience. They also have paperwork to complete and customer logs to keep up-to-date, all within a specific timeframe. They always work indoors, under artificial lighting, seated in the same position for hours at a time and are often exposed to sounds and noise levels which can be both uncomfortable and distracting. While they work in close physical proximity to their co-workers, shift patterns are often configured to minimise the level of contact (and communication) occurring between employees; staggered breaks maintain a sense of isolation which deters drops in productivity and, a cynic might argue, keeps the threat of widespread unionisation at bay. Call centre workers must also maintain a level of consistent precision in their work and appear patient, understanding and enthusiastic while performing repetitive, potentially demoralising tasks (like trying to upsell products to irritated customers or telling someone their bill is overdue) with little autonomy over how they do their jobs and under constant surveillance. This work-related pressure is then exacerbated by the certain knowledge that they are easily replaced if these conditions impact them adversely or, God forbid, they should dare to express a problem with them.

Call centres, and other forms of low-paid and precarious work, dominate the employment horizon for many struggling working-class communities in Britain due to the low entry level and constant recruitment drives, but these are hardly jobs people apply for with dreams of carving out a long-term career. The over-abundance of workers means employers have a steady stream of employees willing to accept the pay and conditions – whether or not they are fair – and this is evidenced in the sector's high levels of staff turnover which, despite the cost of constantly recruiting new staff as others leave, hasn't created a sufficient financial incentive to improve conditions, thus retaining workers for longer periods. The nature of the work is so intense, even the 'good' call centres can be tremendously dispiriting places to work.

'It's very precarious,' said Craig Anderson. 'These jobs can disappear as quickly as they come in.' He explained that the terms of employment can vary, from being paid between assignments – a now outlawed arrangement whereby workers were paid even if they were not on placement but which also exempted them from equal pay to salaried employees – to zero-hours contracts. Why does this sector appear so keen to set up shop in one of the most economically deprived regions in the UK? 'There seems to be an endless abundance of young, predominantly young, labour available.'

I ended by asking Craig what he thought the chances were of a working-class person from Greenock, being promoted to any higher in a company, having started as a call-handler. 'I'm trying to think from my personal experience,' he said, 'but I can't think of anybody who went from the call centre floor up to senior management level.'

The near 20-year-long debate about call centre cultures and conditions – whether they are electronic sweatshops or situated at the vanguard of workplace innovation – often obscures the most pertinent fact concerning this sector: it's an industry which relies not only on an abundance of young working-class labour but also one which must, as an imperative, operate in areas afflicted by historic socioeconomic problems. Let's be under no illusions here: middle-class people would not accept having their toilet breaks timed, and rightly so. Being subject to that kind of unreasonable expectation (while providing goods or services which are regarded by many as 'essential') is the definition of 'working class'.

Why are there so many young people looking for such precarious and often unpleasant work? Why aren't they in further or higher education? And if there is such an abundance of young workers living in poverty that these employers can attract, what does that say about the education system they spent a minimum of 11 years of their lives in?

The biggest predictor that someone will end up in low-paid, insecure work is low educational attainment. The biggest predictor for that is which postcode they grew up in. What many people do not realise (or refuse to accept) is that this relationship between schools in impoverished areas and precarious, poorly paid, often futureless work is no happy accident – it is by design.

Back in August 2020, our attentions turned briefly from the Covid-19 death count to the imminent publication of secondary school exam results. Across the UK, it was decided that in the absence of traditional examinations due to school closures, results would be assigned solely by using an algorithm. The algorithm would be fed information that all teachers were to provide, including their own estimates of pupil grades.

Erin Bleakly was one of thousands eagerly awaiting her results. A grade A student from working-class Shettleston in Glasgow, she declined the offer of receiving results by text and instead chose to receive them by post. 'I was expecting quite good marks,' she recalled, as we enjoyed a seat on a grassy mound overlooking her school, some weeks after the exam scandal which brought the issue of educational inequality into sharp public focus.

'Throughout the year, I'd done consistently well. I'd never failed a test. Never.' Then the envelope hit the floor. 'I remember opening it and being so happy. Then I seen the letter inside.' Four of Erin's results were downgraded. Her childhood dream of becoming a vet suddenly seemed less likely, despite her being on course to pass the exams with flying colours. 'My legs collapsed beneath me as I started wailing, crying. I couldn't believe it. All my hard work for the past two years had now vanished. I got an A in my maths prelim and then I got a D.'

As morning turned to afternoon, social media buzzed with similar stories. Pupils who had been expecting decent results found they had been marked down. Some teachers broke rank, publicly condemning the Scottish Qualifications Authority's algorithm which, until then, few knew anything about. While parents vented their fury at their kids being downgraded, the initial assumption was that some error had occurred – a most serious misapprehension.

The memory was painful for Erin to recall, and no wonder. 'I took a really, really, really bad panic attack that morning to the point I couldn't breathe,' she confided. 'The experience is actually quite haunting to know that something can affect you that much. At first, I thought it was just me, but when I spoke to a lot of my friends, they were the same.'

Within hours a picture began to emerge. A familiar one. Early analysis discovered that the SQA had moderated teacher estimates for pupils from poorer postcodes more harshly than those from affluent areas. Previous pass rates for postcodes – a key indicator of social class – were fed into the system in an attempt to identify and correct for anomalies. This effectively suppressed pupils from less affluent areas by downgrading their teachers' estimates, thus bringing results in line with lower grades historic to those postcodes. While no politician or public official dared speak its name, 'class' was on the tip of a lot of people's tongues. As the backlash gathered pace, coming to the attention of the entire British media, the Scottish government stubbornly dug its heels in, hoping to tough it out, insisting the results were fair. Talk of class war swept the nation.

As well as exposing systemic barriers inherent to education, the scandal also revealed many of the tropes associated with class inequality more generally. Tropes which echo across the socioeconomic landscape and which we will explore in greater depth over the course of our journey. These include: the presumption of merit on the part of those who do well and the lack of analysis of the individual and systemic advantages they enjoy; the notion that hard work and the right attitude have more bearing on outcomes than being from a wealthier family, despite it requiring far greater effort and resilience to succeed when hailing from a poorer background; and the mass delusion, despite a tectonic plate of evidence to the contrary, that the system is, despite its flaws, fundamentally sound – when it is definitionally unfair.

'We were being judged on our postcode or made to feel that our postcode defined us,' said Erin. 'That's when I started the Facebook event.' Within 24 hours, Erin had organised a resistance to the Scottish government and the SQA. A socially distanced protest at George Square in the heart of Glasgow city centre drew modest numbers as well as the glare of the UK's media. Young people held banners aloft that read, 'Judge our work, not our postcode', 'The SQA done us dirty' and 'We have class, you have classism'.

Seven days after Erin's protest, the outcry had only deepened. The Scottish government soon caved, recognising that even the middle-class parents they had banked on looking the other way were furious. There was nowhere left to hide and it was announced that results would be reverted to teacher's predicted grades – as campaigners, including educator and author James McEnaney, had advised months before the results were published.

The justification for the downgrades was, apparently, to preserve the 'integrity' of the system by correcting for teacher bias with respect to final results. Integrity which was ostensibly threatened by the prospect of more working-class pupils than usual receiving higher grades. Take a moment to truly consider the implications here. The higher the number of working-class kids passing with good grades, the more excessively the algorithm intervened to correct for the perceived 'mistake'. For how could a teacher's estimates possibly be construed as anything but over-generous if so many pupils from historically poorer communities performed as well or better than pupils from more affluent ones?

It is in the exams scandal that many of us were, for the first time, given a glimpse of the education system's previously opaque interior. By marking grades down, the Scottish Qualifications Authority would, the Scottish government believed, remain credible – but 'credible' in who's eyes? Upstanding, socially responsible employers like Amazon and Sports Direct? Bed-wetting opinion columnists who haven't worked a day in their lives? Or the countless universities that had only days before progressed thousands of middle-class stoners into the second and third years of their degrees off the back of two months of ropey course work? Given how many rules, regulations and conventions had been thrown out as a result of the pandemic, and how much support was made available to individuals, households and enterprises struggling to cope, it speaks to the integral role educational inequality plays in British society, with respect to the distribution of opportunity, that cutting the poorest kids some slack (and going with teacher estimates as had been recommended) was regarded as a bridge too far.

What should send a shiver up the spine of anyone with a genuine interest in educational equality is how the system's eyes were trained on identifying and eliminating even the slightest advantage from which a poorer pupil might benefit (in this case, being given an unusually high grade) while looking the other way as the myriad social, economic and cultural advantages conferred to pupils from wealthier backgrounds by lottery of birth continued unabated. Advantages which guarantee the most affluent children will snap up almost all the university places and the secure, well-paid employment opportunities that come with them, every year without fail – while precarious employers like call centres stand ready to pounce on the imminently abundant young labour.

Boiled down, the Scottish government's argument was that the many advantages enjoyed by the most affluent pupils – extra tuition, households more conducive to study, lives free of poverty – were not unfair but that the potential for a working-class pupil to receive a B instead of a C represented an existential threat to the entire system of education.

James McEnaney is an author, journalist and college lecturer as well as Scotland's leading educational equality campaigner. His writing and analysis of the education system goes beyond commentary on class sizes and the curriculum and cuts to the very core of the problem – that the system is rigged against poorer pupils. When I spoke to James, he explained:

> 'I suppose the thing people sometimes don't quite grasp is just how embedded this stuff is. Every year, the system ticks over on the foundational assumption that a certain number of students must fail, and then sees no problem with the burden of that failure being disproportionately shouldered by the poorest. Without the algorithm, the SQA could still have found a way to adjust national pass rates so they fit previous years, but it was only worth doing that if they could also bake in the pre-existing socioeconomic disparities, which is why they forced teachers to rank their students and then used a school's past results to determine an acceptable range of grades that young people from those postcodes would be permitted.'

Like Cameron's economic aviation metaphor, the phrase 'exam algorithm', as well as terms like 'attainment gap' (which attempts to account for why some kids can play the oboe by age 12 while others cannot write their own names), obscure a counterintuitive and ugly truth – education systems in Britain are configured not to close the gaps between the affluent and the poor, but to guarantee them.

We inherit our social class at birth, usually from our fathers, whose social position is defined by their occupation, but it's in the education system where class inequality is accelerated and where the historic gulfs between social classes are formally cleaved. Both historically and presently, two-tier state education systems are operated across the UK. Like most Western education systems, Britain's various iterations are rooted in a long-held belief that not all children are created equal. While pupils from poorer backgrounds can and do ascend, and pupils from more affluent backgrounds may do less well than anticipated, the occasional outliers (or invigorating parental anecdotes) have little bearing on a century of children from wealthier backgrounds doing generally better than those from poorer ones. We often discuss this problem in educational attainment as a gap which needs 'closing', when, in truth, this framing diverts us from the education system's central mission – to ensure a baseline of class inequality which meets the needs of the labour market.

While the focus of educational reforms – which have sought to improve educational standards and reduce inequality – is often on what and indeed how children are taught, the disparities between their parents and caregivers, in terms of education, health, housing, employment and income, play a far more decisive role in producing gaps in attainment than discourse suggests. Do you own your own home or are you renting? How much do you earn and is this income stable? Are you on a zero-hours contract, working freelance or are you salaried? Do you have a parent in prison? Do you have adequate childcare? How much of your income do you pay towards accommodation and utilities? Are you a single parent? How much time do you have to spend with your children after work and domestic tasks are completed? These are the factors which largely predict educational outcomes and they are all linked with systemic inequalities.

The postcode parents can afford to live in will largely determine which school their child will go to. The marketisation of the state education sector (where parental choice sets schools in direct competition with one another) naturally favours families who are socially mobile enough to parachute themselves into coveted catchment areas while households in deprived communities must make do with whichever schools are on their doorsteps. The quality of education (and by 'quality', I mean the demands placed on teachers, pupils and resources by the socioeconomic conditions in which a school is situated and not the quality of teaching staff) then varies wildly along lines of social class.

According to the Child Poverty Action Group, there were 4.3 million children living in poverty in the UK in 2019–20. That's 31 per cent of children, or nine in a classroom of 30. There are expected to be 5.2 million children living in poverty in the UK by 2022 – an increase of nearly one million in just two years. While being working class and being poor are not always the same thing, there is a clear divide between those on the lower end of the income scale and those earning decent money.

The government's 2020–1 State of the Nation report on social mobility confirmed that, today in Britain, you are still 60 per cent more likely to be in a professional job if you are from a privileged background rather than a working-class background. While top-line statistics make great headlines (as well as sobering reading), they can often obscure the multifaceted and structural nature of gaps in educational attainment, which are so complex that a working-class kid achieving the same results as a middle-class one does not necessarily mean they will enjoy the same quality of life. There are vast distances between children from advantaged and disadvantaged backgrounds with respect to what their parents earn, how stable this income is and what those earnings afford you on a day-to-day basis. But there are also gaps in the quality of resources available to them. Health services in middle-class areas, for example, are not as stretched, due to lower population density and less instances of emergency and illness, but this is not reflected in how services are funded, meaning middle-class people can expect a higher quality of healthcare than working-class people, including more time with a GP. In housing, poorer families are subject to residential instability as a result of the lack of affordable social housing, which leaves them vulnerable to a precarious rental market, where they may be turfed out with little notice creating educational disruption which may even involve kids having to move schools. Less well-off kids are likelier to have to share bedrooms, which is less conducive to study. Parents are likelier to be working longer hours on irregular shift patterns which may require childcare – another burden on household income – and have less time and energy to read with their kids or help them with their homework. Children from working-class and poor backgrounds are also likelier to be care-experienced – taken into care by the local authority for their own wellbeing or protection – and are also likelier to be performing a care role within their household.

Not even the proliferation of digital technology and the internet has mitigated the distance opening up between advantaged and disadvantaged young people: according to the UK government's own Social Mobility Commission, when the pandemic hit in March 2020, 'only 51 per cent of households earning between £6,000 to £10,000 had home internet access, compared with 99 per cent of households with an income over £40,000 and even when poorer households had access to equipment and internet, they were then less likely to have the skills to utilise it.'

If a child reaching their fullest potential was as simple as having a committed, innovative and passionate teacher, then education systems across the UK would have required nothing like the level of reform to which they've been subjected over the last hundred years. The algorithm used by the SQA to moderate grades initially, revealed that irrespective of the effort and ability of pupils from poorer backgrounds, and indeed the dedication and graft of their teachers who work tirelessly to help to progress them through school, the system itself acts to prevent working-class children from reaching their full potential.

The true determinants of a child's educational success lie in their early years. Before a child even begins to learn formally, their core cognitive capacities may already be largely shaped by their upbringing, itself shaped by the social position of their parents. Among the many advantages enjoyed by children from affluent backgrounds is the likelihood they'll be raised in a dual-income household where the division of parental labour and financial responsibility creates a household environment conducive to a child's early development.

Two parents in secure, well-paid work can afford to plan further than the next week. They may be better able to provide children with access to leisure and cultural experiences. But most decisively, such a child will generally (though obviously not always) be exposed to lower and less prolonged levels of household and community stress than poorer kids. Better health means less illness and fewer bereavements. Better housing means a stronger sense of security and less absence. And better employment means less poverty-related stress and household adversity.

The culmination of these advantages is that by the time a middle-class child enters an educational institution, they often possess a core resilience which better orientates them in the highly standardised educational environment. One which is already carefully calibrated to the learning and behavioural needs of middle-class children.

In contrast, less affluent children are likelier to experience over-exposure to poverty-related stress, which increases the poorer their parents are. Residential instability, poor-quality housing, parental ill-health, bereavements or the presence of a family member with alcohol or substance misuse issues place poorer families under additional strain. Throw in a cut to Universal Credit, or a parent dying suddenly, and the whole notion of a child being 'school ready' by age five becomes unrealistic. And data shows that the attainment gap increases with every stage of education – all the way to university level, if a working-class kid makes it that far.

While we must, I suppose, commend governments for at least acknowledging the poverty-related attainment problem and concede, grudgingly, that the fact we even track it is grounds for a modicum of false hope, annual rhetoric about closing gaps in attainment remains no more than political grandstanding – the same politicians who assure us that more is being done to reduce the distance between middle-class and poor kids are meanwhile further entrenching the broader systemic inequalities in housing, health and the labour market, which lie at its root. As long as poverty on this scale exists, children from poorer backgrounds will fare less well in education than comparatively more affluent kids.

The effects of class inequality are multiplicative: a child is born poor, they experience more adversity, their development may become compromised and they then enter an education system where teaching practices and educational cultures are misaligned with their needs. As a result, upon leaving school, they become likelier to end up in low-paid precarious work and, eventually, rearing their own children in various forms of poverty, and so the cycle repeats.

Pay attention to which areas the 'best' schools tend to be situated in next time performance league tables are published. You will note a correlation of sorts between schools which produce consistently higher grades and postcodes which produce consistently higher house prices – this process mirrors generational poverty in that it is cyclical, but rather than reproducing poverty it reproduces prosperity.

*

That being the case, the question we must then ask ourselves is: what is the mission of a state education system? Despite romantic notions of universal education as a great leveller, it exists today as it did when it was instituted to serve the labour market. There are, after all, only so many well-paid, secure jobs available, just as there is always a minimum of low status, precarious and poorly paid positions needing to be filled. Education systems act as pipelines, feeding the economy with 'skilled' and 'unskilled' workers and professionals. When we understand this, we begin to see the 'abundance of young labour' in places like Greenock in greater context.

In the 1920s, for example, it was the view of the establishment that the working classes – who made up three quarters of the population – required only a basic education. This led to around 1.5 million working-class people finding work as domestic servants – in the households of Britain's expanding upper-middle classes.

Only when we consider education's history as the exclusive preserve of the wealthy can gaps in attainment between social classes be adequately contextualised. Universal education is, after all, a fairly recent development. It was not until the nineteenth century that poorer children were considered worthy of teaching and even then, so-called Ragged Schools, initially organised by volunteers, had no formal premises. Set up in working-class districts in London and then Scotland, teachers held classes in lofts, stables and beneath railway arches. The onset of the Industrial Revolution (and the need for a workforce to power it) is why the British state introduced compulsory education to working-class children in the 1870 Education Act. While educational standards have improved and working-class children are now taught many of the same core subjects as middle-class kids (and cannot be legally denied the opportunities which have historically gone to better-off kids), the education system's relationship to the labour market remains consistent with its historic mission and so children from less well-off backgrounds face an uphill struggle to compete in both education and in work. Educational inequality is effectively imbedded within Britain's wider industrial strategy. The attainment gap persists today not as an anomaly but as an echo of universal education's original mission to provide the labour market with an abundance of workers, willing to perform difficult manual tasks. Schools in more deprived areas face challenges not simply as a result of poverty but also because working-class and poor kids receive less money per head, are exposed to higher levels of staff turnover, higher incidences of temporary supply teachers and often teachers who collectively possess less experience.

Inequality is written into the education system's DNA. The educational offer made to poorer children is not only inferior to that offered to middle-class kids but also subtly restricts their potential by constraining their aspirations as they slowly become adjusted to the idea people like them just don't do certain things in life. When kids don't see their peers succeeding, they learn something profound about their own place in the world. Children attending a school in a poorer community are placed at an immediate disadvantage and it is in the generational reproduction of this structural disadvantage that the education system derives a great deal of its 'credibility'.

Diane Reay is visiting professor of sociology at the London School of Economics and emeritus professor of education at the University of Cambridge. Her 2017 book *Miseducation: Inequality, Education and the Working Classes* draws painstakingly from research as well as hundreds of interviews with children to explode the myth that there is any great difference between the education system today and when it was first instituted. Reay writes:

> 'Historically, the English educational system has educated the different social classes for different functions in society. However, in the 21st century, the expectation is that the English state system is providing roughly the same education for all – it does not. Even within a comprehensive school, when young people are all being educated in the same building, the working classes are still getting less education than the middle classes, just as they had when my father was educated at the beginning of the 20th century. We are still educating different social classes for different functions in society.'

While today, great lip service is paid to closing the gap between the wealthiest and poorest students, our economy would simply seize up were this gulf suddenly eradicated. If young people from poorer communities didn't drop out of school early or fail to achieve high enough grades to go straight to university, then who would do those low-paid, precarious jobs? Who would be there to answer your call about your car insurance at 11pm? Who would be working the drive-throughs when late-night hunger strikes? And who'd be sat on the checkouts in 24-hour convenience stores, petrol and service stations, or staffing the late-night bars and eateries we've come to expect to remain open long after most people are asleep. Who would be the taxi and bus drivers we expect to take us home after a night on the town?

But the poverty-related attainment gap – which is so wide that some fall off the grid after just a few weeks of absence from education while others can afford to take 'gap' years and still not fall behind – also has clear political roots. Inequality, to the extent we see it across Britain, creates perverse political incentives. If the scales were suddenly tipped in the favour of working-class children (higher funding, action on systemic issues like housing and perhaps even reform of how kids are assessed that accounted for various disadvantages) and poorer kids, as a consequence, began mounting a serious challenge to middle- and upper-class dominance of universities and prestigious professions, how do you think economically active, property-owning, electorally lucrative parents – whose expectation that their children do better than poorer kids often borders on entitlement – would respond?

Think of the anger and dismay expressed by many working-class parents (and the tabloids they read) when it was found in June 2021 that children from ethnic minority backgrounds were performing generally better than white ethnic boys. That debate was cynically framed by some politicians and sections of the press as working-class white kids being disadvantaged by a system which was over-correcting for racial inequalities – not as evidence of the merit of BAME kids – and served as a deliberate distraction from the far wider gulf between working-class kids of all ethnic backgrounds and the middle- and upper-class kids peeling away nearer the top. Can you imagine the headlines if middle-class kids suddenly faced serious competition in education and the labour market from people from deprived areas?

Politically, there is little incentive for politicians to confront (even rhetorically, let alone truly address) the true systemic nature of the attainment gap between middle-class and working-class children; it would involve reforms which many middle-class parents might construe as attempts to constrain their offsprings' progress. The political retribution from this powerful electoral demographic in the event a perceived curtailing of their advantages occurred would be nothing short of seismic. Politicians understand this; therefore, the attainment gap is maintained (albeit with some tweaks here and there) as it remains the essential mechanism by which opportunity is distributed upward, not only in accordance with the requirements of the labour market but also with the expectations of upwardly mobile, culturally dominant, electorally pivotal middle-class parents.

The attainment gap is merely one of various outputs arising from a socioeconomic system which is configured to intuitively serve the interests and aspirations of wealthier social classes. In the UK, education is not a meritocracy; it's what sociologist Miriam David termed a 'parentocracy'. When annual league tables of the 'best' schools are published, they are not necessarily a reflection of the abilities, attitudes or aspirations of individual pupils, but of the advantages and disadvantages (and collective political clout) associated with the socioeconomic position of their parents. These advantages are then supplemented by systemic quirks, baked into the education system itself, which interact with similar inequities across health, housing and so on to entrench this class privilege – the exam algorithms being merely the latest method of attempting to ensure this.

Even if we eradicated the attainment gap in state schools tomorrow, and kids from different social backgrounds were placed suddenly on an even playing-field, Britain always has a back-up plan where the entrenched advantages of the wealthy are concerned. An insurance policy which would guarantee, in the unlikely event that Britain ever became a fundamentally fairer society, that the 'best' opportunities still went to the 'best' (most privileged) children – the independent school sector.

Britain's most influential people in politics, in business, the media, Whitehall, public bodies, local government and the creative industries are over five times likelier to have been privately educated than the general population. A 2019 report by the Sutton Trust and the Social Mobility Commission found that just 7 per cent of British people have attended a fee-paying school, yet two fifths of those in top positions were educated privately. It isn't hard to understand why wealthier parents would send their children to these expensive institutions: a private education grants even the most academically unremarkable student an automatic VIP priority-pass into the world of highly paid employment.

But as with the poverty-related attainment gap, the quality of education is not the main issue; it's the structural basis of the various other advantages which must be understood.

As with many inequalities, the pandemic revealed a lot. In Scotland, analysis carried out by Barry Black, a University of Glasgow researcher, later reviewed by Professor Catherine Lido of Glasgow University and then reported by James McEnaney in investigative journalism platform The Ferret, revealed the poorest kids were hit four times harder by grade moderation than better-off kids. Data showed that 'in schools where 40 per cent or more of pupils receive free school meals, the average number of Higher grades moved from pass to fail was more than 20 per cent' and 'in schools with 0–10 per cent of pupils receiving free school meals (an indicator of social class) the figure was 9 per cent'. In private schools, where no children receive free school meals, that figure fell to just 5.1 per cent. This trend was reflected across the UK, though kids in England from the poorest backgrounds were not moderated quite as harshly as kids in Scotland were.

Now let's briefly consider the systemic advantages enjoyed by independent schools and, indeed, what the private school system truly symbolises. The first advantage is how much more money they have: the gap between private school fees and state spending per pupil in England has more than doubled over the past decade, according to the Institute for Fiscal Studies. Average fees, of £13,600, were more than 90 per cent higher than the £7,100 spent on state-school pupils in 2020–1, compared with a gap of 39 per cent in 2009–10. As mentioned, charging fees means poorer kids are filtered out automatically, which includes kids with complex behavioural or learning needs commonly associated with areas of disadvantage. Put alongside smaller class sizes, this simplifies the teaching process and educational culture significantly.

But private schools don't just benefit kids, they also, potentially, benefit their parents, too. As well as already being on secure incomes high enough to send their kids to fee-paying schools, the more they earn, the more parents can take advantage of certain perks which become available to people in that income range. For example, most independent schools offer 'fees-in-advance' schemes. These require a substantial lump sum of tens-of-thousands at minimum and leverage the independent school's charitable status, turning it into a little on-shore tax haven. The wealthiest parents can pay their children's school fees up-front – anything from a single term to several consecutive years of private education – and while this initial outlay is substantial, the school can then plough the capital into low-risk investments, the returns from which are not subject to taxation. The school and the parents can then split the windfall. How convenient.

This is just one of a variety of ways the wealthiest families' kids not only stand to benefit, but also the parents. As reported by financial magazine *This Is Money*, there are all sorts of legal loopholes that wealthy people with access to accountants and financial advisors can exploit. Parents can use offshore bonds or tax-free pension pay-outs. Grandparents can set up companies and make their grandkids share-holders, paying their school fees through tax-free dividends. Private schooling is not simply about parents giving their kids the best opportunities, it can also be about parents advantaging and enrichening themselves – that is part of the product being sold in the independent school system.

This extreme exclusivity and utter insulation from the real world are a great draw for many. As well as an enhanced educational experience for children, they (and their parents) are also plugged into historic and lucrative social and professional pipelines where paths to progression are less obstructed than those from the state sector and where intertwining social networks engender cultures of nepotism and cronyism which find expression later at the highest levels of business and governance as sleaze and corruption scandals. Families become highly networked within these prestigious institutions and, as a result, opportunities become less a matter of merit but materialise from simply picking up the phone or taking someone for lunch. In these circumstances, an academically unremarkable boy may rise to the very apex of British society, despite being of very low character or ability, as a result of the resilient social connections, a sense of entitlement and relative protection from serious accountability that a fee-paying education can provide.

But by far the most significant factor arising from the independent sector (especially from the super-elite schools like Eton) is that it virtually guarantees that British society will always, to some extent, be ruled, across its main institutions, by a privately educated elite. In a best-case scenario, these privileged individuals, despite their remoteness from the majority of the population and its problems, will act on sincerely held values and beliefs, and possess the necessary humility and open-mindedness to learn what they do not already know about the experiences of others in society. In a worst-case scenario, a significant number of these privileged individuals will harbour fixed beliefs, especially about the root causes of poverty and their own wealth, as well as delusions pertaining to their talents and cavalier emotional attitudes rooted in an under-exposure to serious consequences for self-seeking or unethical behaviour – engendering an unwarranted sense of overconfidence in their insights and abilities.

All social classes contain a great diversity of opinion, values, aspiration, political beliefs, abilities and expertise. Where the greater imbalance in Britain is concerned, is that for all that diversity at the very top, even those who appear diametrically opposed in terms of their personality types and politics, share the same basic class interests. It is those interests – the main one being a deep personal and professional investment in the preservation of the status quo – that will subtly motivate their reasoning of the various problems arising in the areas for which they inevitably become responsible in public life and business. This near uniformity among elites in that one area – class interest – and the disproportionate level of power, influence and prestige held by this group go some way to explaining why Britain remains wedded to the same prevailing economic orthodoxies that create systemic social inequality as a toxic by-product.

The continued presence of private education across Britain underlines the deep commitment of governments across the home nations to the reproduction of class disparity and all of the socioeconomic mayhem this entails for working-class people, the poor and the vulnerable, despite private education being nothing short of a public catastrophe for social equality the country over.

*

Aside from the clear systemic inequalities in education, there is a far graver moral issue arising from a system that is gamed in favour of those who already have enough: the poorest, most disadvantaged kids suffer as a direct consequence of wealth and opportunity being hoarded by the dominant social groups. Perhaps the starkest warning that systems of education require a fundamental overhaul is the worryingly high number of children who are failed by them. In England, it is estimated by Ofsted that between 200,000 and a quarter of a million kids are being taught in schools which are underperforming. In 2019, research for the then-children's commissioner for England found that around 100,000 pupils leave school every year without basic qualifications (a 28 per cent increase since 2015). This may be an indication that educational cultures – shaped from a distance by politicians with short-term political objectives – have played a pivotal role in the educational exclusion of millions of children who, as a result of prolonged, poverty-related adversity in their early years, could not be accommodated by schools. Kids who may be so traumatised by poverty-related neglect or abuse that they often have more in common with war veterans than they do with their social workers.

Education is about more than teaching. It comprises many social, cultural, relational and financial factors which inform and shape pupil experiences. While much emphasis is understandably placed on pedagogical practices, curricula and class sizes, the quality of a child's education is largely determined by factors outside the classroom, and which usually precede their enrolment in school. However, the most disadvantaged children are disadvantaged not only by their circumstances but also by educational cultures which are poorly adapted to accommodate their unmet needs. Unmet needs which increase by orders of magnitude the higher the level of deprivation a child experiences.

The likelihood of a troubled child remaining engaged in their education depends almost exclusively on their ability to behave in a manner expected of them. A manner their teachers have been trained to accommodate. This behavioural expectation underpins mainstream educational cultures.

All children are prone to occasionally challenging behaviour. The question where some of the most vulnerable kids are concerned is whether their behaviour is genuinely challenging or whether too many teachers simply deem it challenging because they are personally and professionally unequipped to deal with it – there's a difference. How many kids were caned as a result of a teacher's unmanageable stress levels or desire to engage in power-plays? Children flagged up as 'bad apples' are often, contrary to popular belief, extremely socially sophisticated and tremendously resilient. The problem is that their emotional intelligence and capacity for resilience are marshalled by the necessity to navigate hostile social environments at home and in their communities. Place these young people in an environment most other people would find frightening and they will often seem at ease. But place them in the classroom of a mainstream school and they'll become irritable, defensive and adversarial.

I often use the analogy of Peter Parker, the secret identity of Spider-Man, when he is first bitten by a genetically engineered spider, which imbues him with special abilities including enhanced speed, strength and agility, as well as an ability to scale vertical surfaces like a spider. But his greatest skill, one which gives him the decisive advantage, is what Parker refers to as his 'spider sense', which 'tingles' whenever danger is afoot. He often detects threats before they occur and takes evasive action to avoid injury and death. Hyper-vigilance, as experienced by those who have endured neglect and abuse, works in much the same way: victims develop an instinct for avoiding or mitigating harm. But take them out of the hostile environment and they become unable to regulate their spider sense. Imagine you possessed such a sense but it was poorly calibrated. Imagine it tingled when someone dropped a plate in a café or whenever you heard a raised voice or police siren. Imagine it tingled at the wrong things and filled you with fear, dread or anxiety unnecessarily. Do you think you'd become a 'friendly neighbourhood Spider-Man' or an isolated, paranoid anti-hero?

Education selects for those children who possess the capacity, first and foremost, to emotionally regulate, to ask for help, to state their needs clearly and to ritualistically retain and then recount information with a degree of confidence. Those children who fail to hit the traditional milestones as outlined by the curriculum, and as a result become increasingly excluded, may be regarded as intellectually inferior, whereas often they have simply developed different strengths and aptitudes which the mainstream education system does not value and therefore does not nurture.

When we talk about poverty, and we ask why some children do better than others, we must confront the reality that disparities between children and young people from advantaged and disadvantaged backgrounds are not always a question of merit, effort or resilience but a matter of neuroscience. Differences are then compounded by educational cultures that favour children who conform to behavioural and learning expectations. This is a difficult line of enquiry as there is a risk that by acknowledging differences in brain development between different cohorts of children, some will take it as a green light to insist that educational inequality, to the extent that it exists in Britain, is inevitable – it isn't. We must view these differences not as innate but as responses to various forms of stress in a child's early years. Persistent stress, often referred to as 'toxic stress', alters the architecture of a developing brain. Neurocircuits which have become damaged, or which have been under-stimulated, require intervention and enrichment to repair or develop. This requires greater autonomy for teachers on the frontline, trauma-informed training and a re-examination of 'punishment' as a concept, generally.

Under the duress of hyper-vigilance, a child will struggle to respond positively to the traditional social cues and behavioural incentives found in a classroom and this becomes the basis of their incremental exclusion – not simply from school but from society. A teacher who does not understand how trauma works may, for example, enter a child's personal space in order to exert their authority but when that child lashes out and strikes the teacher instinctively, it is the child who will be sent off to the 'special school'. Effectively, these children risk being excluded from mainstream education not due to a disability or grave emotional disorder, but as a result of the naturally occurring adaptations they have undergone to successfully navigate unreliable, potentially dangerous adults. When teachers resort to passive aggression in the classroom (as they often do), they trigger security instincts in children where trauma is present – then punish them for reacting naturally.

Now consider how such a child's progress is tracked in a mainstream education setting and why this aspect of educational culture sets many up for failure. There are objective measurements such as tests, assessments and, eventually, examinations. Then there is the behavioural dimension of a child's progress. What is their attitude to work, how actively do they listen, how well do they interact with others and, ultimately, how compliant are they? Whether objectively assessed or behaviourally interpreted, the child is always being measured against the hypothetical standardised child. An ideal specimen against which some children will fare better than others. The central problem with mainstream teaching is that it cannot adequately accommodate the neurological reality that every child's developmental trajectory is unique. Some hit developmental milestones sooner than others. Some require additional support to 'keep up'. And others face challenges at home which often present in the classroom as 'challenging behaviour'. There are gaps between boys and girls, affluent and poor, non-disabled and disabled, and neurotypical and neurodivergent. Yet children of the same age are generally compelled to sit exams and tests at the same time.

While great efforts have been made to accommodate the learning diversity present within mainstream education, children are nonetheless still subjected to a regime of testing and behavioural expectation which disadvantages those children who develop unconventionally or at a different pace than the 'standard'. And the overwhelming majority of children who require more time, as well as more dynamic, innovating teaching techniques, come from poorer backgrounds.

Despite everything we now understand about early years development, our education system is poorly equipped to identify how it provokes the very behaviours it punishes and clings desperately to teaching methods largely rooted in the early twentieth century – belts have just been replaced with time-outs, naughty steps and shame culture. Traumatised children don't need to be punished, humiliated or shamed, what they require is patience, love and encouragement.

In an education system in which corporal punishment was only recently outlawed, is it any wonder that so many children from poorer backgrounds, dealing with trauma and adversity, struggle to succeed within educational cultures which remain, in many ways, underpinned by the same tough-love philosophies that resulted in the routine belting, caning and humiliation of children? Violence was a cornerstone of education across the UK until fairly recently, despite debate about the efficacy of corporal punishment stretching as far back as ancient Athens. Today we balk at the thought of children being assaulted by their teachers but for most of the twentieth century it seemed as normal as the publication of school league tables, standardised testing and harsh examination regimes do today. There are, therefore, solid reasons to ponder which educational cultures in operation presently we may one day look back on with a sense of bewilderment, regret or even shame.

With all of this firmly in mind, we return to the basic mission of education – to prepare children for the world of work. In recent years, governments have attempted to gloss over class inequality – increasingly driven by government policy – by broadening the parameters of what is deemed 'success'. This represents a concerted effort, on the part of administrations across the UK, to present the mirage of social progress while simultaneously preserving, and even expanding, the longstanding socioeconomic inequalities that produce educational inequality in the first place. Government ministers can cite the number of pupils reaching 'positive destinations' without going into any detail that these include exploitative, low-paid and precarious jobs.

Today, we have a labour market in constant flux where the whole concept of a job for life will soon be consigned to history, requiring millions to train and retrain throughout their working lives. Not only is education often misaligned with the learning and behavioural needs of many pupils but it is increasingly at odds with the labour market it serves, the demands of which seemingly change on a daily basis, driven by technology, where working-class kids stand on the precipice of yet another industrial revolution beyond which their labour may become surplus to requirements indefinitely.

In a two-tier education system, designed at the height of the industrial era when it was very clear which type of work a young person would eventually move into, the uncertainty of what awaits them in the world of work places kids from poorer backgrounds under additional strain, compounded by the systemic and cultural factors outlined in this chapter. Indeed, by the time many of them finish school, aspects of the curriculum may be rendered obsolete entirely by technological advances, as may the industrial sectors they were educated to become employed in. Without a clear sense of where they are going, education simply engenders in children the same sense of alienation many of their parents experience in precarious jobs.

In contrast, children from more affluent backgrounds, less likely to be dealing with trauma, developmental complications and additional support needs, and more likely to grow up in households and community environments conducive to emotional stability and study, enjoying increased access to extra-curricular activities, tuition and higher levels of parental engagement with their studies, stand a far greater chance of moving seamlessly from education into professions which are less precarious, better paid and not as vulnerable to disruption.

When work doesn't lift you out of poverty, and your postcode has as much bearing on your exam results as your effort or ability, yet political leaders still insist otherwise, we should expect more of the same. The gaps in employment and education we see across the UK between social classes are about much more than schools or the labour market; they are two inevitable outcomes – red lights flashing on the dashboard – which point to Britain's profound and unsustainable socioeconomic and cultural imbalance.

No parent likes the idea that their child's educational destiny may be a foregone conclusion. Parents from poorer backgrounds understandably resist the notion their children's potential may be limited by their economic circumstances. Parents from more affluent stock understandably push back against the suggestion their children's success may have as much to do with the myriad advantages they enjoy as it has with comforting notions of merit. It is in these parallel refusals to accept the truth – wealthier kids will always do generally better than poorer ones in education and work – and the self-serving refusal of governments to articulate honestly and confront the structural nature of the problem that we muddle on, in deep collective denial, unable to see education and the labour market for what they truly are – the primary guarantors underwriting Britain's shameful class inequality.

3

Arrested Development
Poverty and prison: a love story

Sometimes you can just smell a horrendously shitty night. Edinburgh's Pilton on a Friday at dusk is no fun, let me tell you. Not if you are sober and north of 16 years old. But that's where I am, looking as conspicuous as ever, trying to deliver what should have been a simple piece to camera in the middle of a long shooting schedule filming my first television series, *Darren McGarvey's Scotland*. My embattled focus is pulled in every direction but the job at hand. We are racing against time, chasing what remains of the natural light, and my inability to deliver the monologue (which is short but laden with complex statistics) is not helping. It's just gone dinner time and we'll be out until at least 3am.

The mild dread I experience is a result of knowing something of what I've just walked into. In spring 2018, in a bid to crack down on what was referred to, at the time, as a 'reign of terror', local police charged 30 men with 133 offences and handed down 20 antisocial behaviour orders. While youth disorder and drugs are problems commonly associated with Edinburgh's housing schemes, the culprits have also managed to carve out for themselves a rather original reputation as motor-vehicle aficionados – other people's motor vehicles that is. For years now, this part of Edinburgh has become infamous for the stolen car and motorcycle culture which involves young people 'acquiring' premium, stylish and powerful vehicles before embarking on dangerous thrill-rides around the community. For them, this is sport. For residents, it's a dangerous nuisance. But for the police, it's a crime wave that never recedes.

Despite my keen awareness that press coverage of this area – and areas like it – almost always errs on the side of sensationalism, I am unable to completely flush the media-generated prejudices I have sub-consciously acquired from my mind. I do not feel safe and, to be honest, could do with something to take the edge off as the final remnants of daylight vanish, bathing the community in menacing darkness.

At the top of the hill, just behind a solitary local community centre, a group of around 30 young people move slowly and obtrusively towards the venue. In all my years of living and working in communities like Pilton, I have never seen anything like it. There is something ceremonial about their formation as they draw closer to the building. But just as I begin to suspect this may be the group of young people we are here to film, they turn sharply and proceed right past.

It's in these moments when I sense danger that the chest is puffed out a little further than is sensible. Thankfully, the youths are far too absorbed in each other to detect my presence, let alone my anxiety. My heightened sense of awareness is a stress response, triggered by the same fear which encourages young people to move around in large, often intimidating packs. The difference, however, is that my most terrifying experience is far behind me; every social, economic and cultural current shaping their attitudes, beliefs and experiences is pushing many of these young people towards theirs.

We head towards the community centre entrance. I am extremely keen to get indoors but the building is locked. Thankfully, after a quick phone call, two friendly looking figures emerge from an office to let us in. James Riordan and Katie Glover are youth workers, run-ning an Alternatives to Crime initiative which aims to draw young people at risk of entering the criminal justice system away from trouble by offering them diversionary activities. Tonight, I will accom-pany James, Katie and a group of young men to a go-karting session and then, after some food, drive with them from Pilton for a midnight stroll on the breath-taking sands of Scotland's east coast – a most unconventional way to spend your Friday night. The official aim is to remove the young men from the social environments and networks which make offending behaviour likelier, but I suspect the unofficial aim is to get them as far away from the over-bearing, clumsy and often excessive local police presence which has become the bane of these young men's lives.

The centre is well-heated and abundantly equipped – two signs it stands a chance of surviving. Community venues are often left to fall into disrepair and they become unpleasant places to spend time. At local authority level, where communities are run increasingly along supply-and-demand lines, the decline in popularity of a public space often registers as budgetary inefficiency rather than a warning to invest. The pervasion of this economic ideology through every level of governance is why warm, well-kitted out, safe community spaces are now so rare. As well-stocked as it is, however, 30 kids just walked directly past it, indicating that even in the presence of long-term resources, and more competent authorities, the challenges facing James and Kate would still be formidable.

Youth workers are dependent on funding from local and national governments and their networks of arm's-length quangos to stay afloat. This dependency creates a lop-sided dynamic in which those who understand best the nature of the challenges faced by the community must justify their work in terms which satisfy the institutional interests of those who govern it poorly, from a distance. Individuals and small organisations, which tend to work most effectively in places like Pilton, must occasionally feign deference to those upstream who control the purse strings – the true definition of 'political correctness' before the term was appropriated as culture war click-bait. Projects of this size can usually expect to be funded for a maximum of 12 months, making it very hard for youth workers to plan long term, both in terms of their own lives and the broader aims of the communities they serve.

Areas like Pilton have been so poorly served for so long that little expectation of better treatment by their elected representatives – or the various public and private stakeholders who control the community from behind an opaque curtain of jargon and bureaucracy – exists anymore. Like investment in local amenities, meaningful political action operates on a strict supply-and-demand basis. Sadly, this is something working-class kids are not taught in school and it shows in their apathy. Perhaps most tragically, they have no desire to resist, believing the obscenely deprived condition of their community, starved of cultural, economic and educational opportunity, is 'normal'.

Many young people in the area are fascinated by motorbikes – it's part of their culture. The problem is that they often lack the resources to acquire them legally. When they are able to cobble together enough to purchase one, the next barrier they face is that there's nowhere to take them for a spin. They are priced out of driving and motorcycling schools, and unwelcome in many other facilities, which leads to the adoption of hostile attitudes towards a community they feel misunderstands and shuns them, making it easier to be pains in everyone's arse. When you see a young man zooming around at high speed on such a powerful vehicle, the tendency is to assume he is misbehaving and that he is reckless and aggressive. He may very well be. But there's also a high chance that he is not only a skilled motorcyclist but one who understands his vehicle intimately, tending to it with the care and pride of an apprentice mechanic.

'There's a family we go to, right, and one of the wee lads is only nine, and he's come out with this motorbike,' James tells me, with such obvious enthusiasm for relaying the story that he barely pauses for breath. 'He's like, "Can you give me a push and try and start it?", and I'm like, "What have you done?" He'd never had an engine and a petrol tank on it, so he'd made an engine and a petrol tank – at nine years old.' I remark on the level of initiative shown by the child in an area where high levels of poverty and poor educational attainment are so often attributed to a 'poverty of aspiration'.

The teenagers James and Katie work with have either come into contact with the police or are deemed at high risk of offending. This often means they're already excluded from using other local facilities – James and Katie are the last-chance saloon for them where access to local youth services is concerned. 'There's other young people that can access everything in the area,' says Katie. 'These lads can't access anything. Nobody's giving them the time of day, nobody's giving them love that helps the boys to think, "Oh, we're part of the community as well."' It is highly unusual to hear a youth worker speak of 'love' in this context, but on the rare occasions they do, it demonstrates the deepest possible understanding of socially excluded youths, their experiences of marginalisation, the values and beliefs which form as a result and how this shapes their emotional attitudes and subsequent behaviour.

Over the course of the hour, the boys arrive and we get acquainted, before taking them to the safest place they can ever truly be – as far from Pilton as is financially possible. Around three hours later, after an exhilarating trip to the local go-karting track, we disembark from the packed minibus into the dead of night. Beneath a clear, moonlit sky, the boys vanish into the darkness. It's a relief to be out of the bus; as sympathetic as I am to the plight of these young people, they are brain damage when contained in small spaces and with unfettered access to stereo sound-systems. The plan is to persuade them to be interviewed on camera but within moments of arriving, their participation appears highly unlikely.

I pull on a coat and survey what little of the surroundings are visible. In the near distance, the boys can be heard shouting, chasing each other, the undergrowth of local woodland rustling and breaking beneath their feet. In this more natural environment, their behaviours do not register quite so threateningly as they might in a densely populated urban setting. Here, they do not cut menacing figures – they are just young boys, playing.

James and Katie lead the way. When the boys show little interest in venturing too far, the youth workers improvise and decide to build a small fire in a clearing in the woods. 'We bring them to places like this and they just get to charge about like mad,' James explains.

'Don't break any fences or that,' he suggests, as the boys test themselves by traversing the natural environment at a characteristically high-octane pace. 'I'd rather they were risking themselves on something like that than risking themselves on a motorbike at 150 miles an hour.

'You'd see them,' he continues, 'and you'd think they're the most confident brash wee guys that ever existed, and they'll shout at you, they'll do whatever else and that, but I say bring them to somewhere like this and it just . . .' his eyes mist over. 'My God, you get to see they're just wee laddies.'

The boys have slowly returned from their adventure and are now situated near or around the fire, albeit temporarily. It's easy to forget that for most of human history, young men, hormonally and culturally primed for strenuous physical activity, spent most of their time in the open air, or in wilderness. Most young people have an energy surplus which must be expended. How that energy expenditure is achieved is often a simple matter of the resources and opportunities available in the environments in which they find themselves. Some boys enjoy days curated according to their interests, passions and needs. Others exist in communities where encouragement is in short supply and where pursuits of interest to them are either out of economic reach or are perceived, by wider society, as 'vulgar', 'dangerous' or 'violent'.

The fire crackles and spits as I move in for a rare heat. One young man has broken away from the group and is stood to my right. I try to strike up a conversation but struggle to hear him properly as most of his face is covered to mask his identity. Many of the boys have prior convictions or are under constant surveillance by police and other authorities. Understandably, they are not keen to talk or be seen on film. It is partly because young men of this type are so reticent to communicate beyond their trusted peer groups that it is wrongly assumed they are either poor communicators or that they have nothing to say. The wild, colourful but playful chatter in the background, away from the cameras and the adults, suggests otherwise.

'You come in earlier on and you said to me,' James says to the masked-up boy, '"I can walk through Pilton now and I don't worry . . ."' so what did you say?'

The boy replies in a deep monotone, 'I didn't get looked at.'

'Exactly,' James replies. 'What was happening before?' he asks.

'Every day,' the boy replies.

Youth workers like James and Katie are far more than chaperones, organising activities for these boys – they are emotional and cultural translators who understand young people intimately due to the proximity they have to them. This can give the impression, occasionally, that they are putting words in the mouths of young people. In truth, they are simply supporting them in the immense task of authentically conveying their feelings and experiences. 'Every single day you got pulled even though you done nothing,' James says.

I venture a question to the boy, asking him what it's like to walk around his own community with anxiety – not about the other youths who are statistically likelier to get into violent conflict with him, but about the police. 'They've got to be all fucking nice to you but not to us,' he says, his voice now brimming with emotion. 'It doesn't matter what the fuck we've done.' Note he does not refer to himself, but to his community. His peers. His tribe. And they are bonded, not by the achieved identity of a sporting activity or the ascribed identity of an educational institution, but by the certainty that police in Pilton exist, primarily, to agitate, disrupt and victimise them.

These boys fit a profile. They are young. They wear hoodies and baseball caps. They drink and smoke in local parks. And they are all, without exception, living in conditions of deprivation. Taken together, this 'profile' is a cocktail of social and behavioural ingredients which virtually guarantee they will be pursued by law enforcement irrespective of the legality of their behaviour. And when that inevitable interaction occurs, these boys become likelier (if they have not already) to progress soon after into a young offenders' institution – prison for children from the poorest backgrounds.

Prisons, by their very nature, are dangerous places. Some more dangerous than others. But you take your life in your hands rising to aggression in a young offenders' institution. Then again, an attempt to pacify aggression can be equally risky. It is often more in a young man's interest to fight and lose badly than it is to be seen backing down.

Emotionally, young men between the ages of 16 and 20 can be immature at the best of times, to the point of appearing childish. This is especially true of young men in prison. But what these boys may lack (or feel compelled to feign they lack) in emotional intelligence, compassion and sensitivity, they overcompensate for with their palpable and imposing presence. Physically, even the unathletic males are in the peak condition of their lives. I once heard an expert at a violence reduction conference compare young men to elite sports cars: when pumped full of the wrong fuel, they become dangerously unstable. Unable to reliably wield their tremendous physical power, they swerve and smash through everything in their path, crashing and burning, leaving anyone in their vicinities impaled by the emotional debris.

It's the spring of 2016. I stand in the centre of a performing arts space, alone, nervously preparing to welcome a small group of young offenders to the first of a series of rap workshops I'm running in Polmont Young Offenders Institution. It's here that many young people from areas like Pilton will get their first taste of incarceration, following interactions with police which, as a result of their social class, are all but inevitable.

Three young men arrive promptly, sizing me up as I invite them to take their seats. Negotiating entry into this volatile peer system requires great care, patience and humility. You must demonstrate strength without initiating force. You must command respect without invoking the blunt language of authority. You must reveal enough of yourself that the participants can safely suppress their scepticism and justify to themselves that you are worthy of their trust. Where expressions of masculinity are toxic, a careful balance must be struck between modelling better behaviour and respecting that the social toxicity for which these boys are famed is often the result of emotional, attitudinal and behavioural adaptations undergone unconsciously in their early years, usually in response to acute episodes or prolonged periods of familial dysfunction and community hostility. Co-sign their bullshit too eagerly, in a servile bid to ingratiate yourself as 'one of the boys', and you'll get found out. But disillusion them too quickly, by pointing out the absurdity of their attitudes and behaviour, naively hoping to reason them out of beliefs they were not reasoned into, and you will threaten them.

It is when these men perceive a threat to their person, their standing in a peer group or their position in a hierarchy that they become likelier to lash out. Understanding and traversing these challenges are far easier if you understand viscerally something of their experience. Having grown up in the kind of community from which these institutions draw almost the entirety of their populations, I possess a cultural frame of reference which a university-educated arts worker may not. I've been on benefits, I've been punched in the face, I've had knives pulled on me and I've seen someone get shot.

The extreme nature of their experiences of violence and crime means these young men operate in a parallel moral world. One which it is possible to understand and even respect without necessarily condoning. It helps, for example, if you have had dealings with the police yourself, even if they were relatively minor. It won't harm your chances either if you have, at some stage in your life, sampled the pleasant effects of alcohol or drugs – or better yet, got yourself into bother because of them. Ultimately, the more your own experience aligns with theirs, the more tools you'll have in your repertoire for successfully traversing the communication challenges which arise when your worlds collide. Common experiences allow for less stilted interactions, providing vital connective tissue; the speech styles, body language, values, and beliefs these boys hold do not appear as shocking, unsettling or vulgar if you understand something of what they have been through. Some time to get to know them also helps but on this occasion, that is a luxury I do not have.

They aim to test me from the outset, in a coordinated manner indicating premeditation. Word has gotten around that I would be visiting the prison and some had prior knowledge of my background as a rap artist and battle emcee. In hip hop culture, 'the battle' functions primarily as an alternative to violence. It's an arena designed specifically to host non-physical conflict – a safe space, if you will. When young men show interest in my status as a battle rapper, I am usually encouraged. You fair better in the prison dynamic if the participants have reason to look up to you. But this time is different. They don't want to hear the war stories. They are not interested in how I got started at the tender age of 19 before quickly garnering a reputation as a vicious, resourceful and dynamic competitor. Instead, they want to see what I can do. They want to challenge me. They desire conflict. It is thanks to my relative proximity to them, socially, and our shared cultural experiences that I stand a better chance of navigating this conflict successfully than a middle-class, university-trained arts facilitator with no criminal record who doesn't believe rap is real music.

Prior to my arrival, the boys had been goading the most talented rapper in their group to face me in a verbal joust. Unable to resist the pressure to rise to a challenge, he is sporting his finest game-face, but I am not in a gaming mood. I feel their eyes scanning mine for signs of doubt or distress. As a facilitator, this encounter is high-risk for a few reasons. This is my first time working in the prison since I was relieved of my duties four years previously due to an alcoholic relapse where I became unable to do my job. So unwell, in fact, that I accidentally carried drugs into the prison, though luckily for me I was not searched thoroughly enough at security. My central preoccupation this morning is redeeming myself in the eyes of a former employer.

But this concern is demoted as my focus shifts to the implications of either side-stepping the challenge to battle this gallus young man or accepting it. If I decline the battle, backing down from the conflict, I may avert a flashpoint with my would-be assailant. That would suit me. Already at a competitive disadvantage given my lack of preparedness as well as being woefully out of practice, I feel in my gut that sense of dread you experience when you must enter the arena despite the certainty you will lose. Then there is the fact I have no prior knowledge of any of these boys. No insight into their backgrounds. No measure of their behavioural triggers or emotional glass ceilings. What if I deploy a 'mum joke' – a form of casual misogyny standard in rap battles – and inadvertently offend? Given the only emotions these men are permitted to frequently express in this repressive environment are apathy and aggression, this wouldn't be the kind of incident you'd be able to talk yourself out of.

This scenario is one that most arts facilitators will never experience. One that most rappers will never experience. There is no protocol nor precedent for how to respond. All I have are my instincts and my experience. Surveying the scene with which I have been presented, I assess that I have two options: decline the battle and potentially squander the opportunity to establish myself quickly with the peer group or throw caution to the wind, embrace the risk and trust that in any event, I'll cope. I hand him the microphone and invite him to take the stage.

Naturally, I let him go first – not to be polite but take his measure. It is immediately apparent that he is a talented emcee in how he holds the microphone. However, it is equally obvious he is riddled with self-doubt – I am too, which is why I am able to detect it in him. Experience has simply endowed me with the ability to conceal tells of fear. Usually, before a battle, I would pace the stage, stalking my prey. I would whip myself up into a frenzy and sometimes even stamp on the floor as if beating a war drum. That strategy is unwise here.

He begins his verse, laying into me for my red hair and bringing up previous battles in which I 'choked' – the unfortunate phenomenon of forgetting what you are going to say and seizing up on stage. It's a sore one for me but the humiliation ignites a desire to retaliate. He raps, while glancing over to me, not to intimidate, but for my approval. It is a rare privilege to glimpse his vulnerability. His eyes have betrayed him. He is frightened – and there for the taking. I act to exploit his anxiety by holding his gaze until he looks away. He finishes his verse and passes the mic. Having tipped the psychological scales in my favour, I sense my gamble paying off and proceed to gently dispose of him in full view of his peers. There will be no round two. He yields without protest. We shake hands and I immediately compliment his formidable skills before quickly spinning the discussion away from conflict towards creativity and collaboration.

It's always a tremendous honour to talk to young men about rap. For many of them, rap, grime and drill music represent their first literary experiences. It's safer for them to engage with language and poetry through aggressive urban music, as this allows them to explore their artistic impulses without looking 'soft'. They are reticent to display sensitivity because their experience tells them that this places them in a position where they may be drawn into conflict. Like the bars on their windows and the latches on their cell doors, their relationship with language often locks them into a fixed and narrow range of expressions and speech styles. This often presents as illiteracy or as poor communication skills. In truth, it's often about self-preservation and deriving a sense of security. While the barriers they face are often structural and lie beyond their control, they are also prisoners of their own fear and low self-esteem.

There's much you can miss if you go into a prison with a narrow conception of the people you'll work with. Among the younger age groups there is a bravado constantly at play which makes it difficult to locate what they do and do not care about. They may appear insensitive and dispassionate but it's all for show – these boys are often deeply wounded and severely traumatised.

There are some safe assumptions you could make, however. You can assume most perpetrators of violence are also victims of violence themselves. That the violence they suffered likely preceded their criminality. And that they were under the influence of alcohol or drugs when the crime that got them locked up was perpetrated. You can also safely assume that pre-criminal adversities they suffered were not perpetrated simply by bad parents or by dangerous peer groups, but also by institutions of health, education, law enforcement and criminal justice which, like the classrooms many of them were excluded from, are not designed to accommodate their complex, unmet needs.

Throughout the history of criminal justice, a tension has existed between three central concepts where punishment by custodial sentence is concerned – retribution, deterrence and rehabilitation. Some are locked up because the risk to the community or society exceeds whatever concerns may exist about the severity of their punishment. This form of retribution is referred to as pure cost benefit and applies to serious and serial violent and sexual offenders whose propensity to harm others is simply too great for a reduced punishment to be considered. Deterrence is when punishments are determined not simply by a cost-benefit analysis but are also informed by the wider aim of making powerful, unequivocal examples to the wider community or society of the consequences associated with certain types of criminal behaviour and activity. In Saudi Arabia, the punishment for drug smuggling is execution by public beheading. While rightly regarded by the global community as extreme, the result is a country with a low reported rate of drug smuggling. But there are also countless examples of deterrence not working, such as in the United States, where the murder rate remains high, or in Scotland, where drug-related offending and deaths continue to rise – despite Scottish prisons being full of drug addicts.

In liberal democracies, where the human rights of prisoners must also be considered, deterrence becomes wickedly complex. A careful balance must be struck between making a powerful example of the consequences of certain criminal behaviour while also allowing for the possibility of a criminal's reform and subsequent return to society. It is here that the third central concept of prison arises – rehabilitation.

Rehabilitation in criminal justice is, essentially, the institutional expression of a belief that people can change. That they can put their past behind them, following punishment. In Britain, there is great confusion about rehabilitation, evidenced frequently in news coverage portraying the 'easy' life people allegedly have behind bars. Reports detail what cons had for Christmas dinner, or the fact some of them have televisions in their cells or pool tables in their halls. What this body of public opinion remains painfully distant from is the fact that people cannot be rehabilitated if they are treated like animals. Prison conditions must be balanced to ensure both that punishment is served and that the desire and capacity to grow and change are cultivated and nurtured. We often forget that for most criminals, the loss of liberty is punishment enough.

Custodial sentences, of course, have a place in modern society, and there are plenty of cases where a custodial sentence or the promise of one have produced positive individual and social outcomes. Victims of crime must also be paramount where punishments are considered. When you become immersed in the harsh and oppressive reality of prison life, from the perspective of the offender, such as I do when I run rap workshops, you risk growing distant from the human consequences of their actions. The bereaved families, the permanently disfigured, the traumatised women and children – it's easy to forget about them when your job is to form a relationship with their abusers and perpetrators.

But where reformable criminals are concerned, particularly young offenders from impoverished backgrounds who are victims of childhood abuse and neglect, key criminal justice concepts such as deterrence and rehabilitation, as practised in the UK, are often deployed counterproductively and require thorough re-examination. They are rooted in assumptions which betray the narrow social experiences of law makers and speak more to the perverse incentives of political leaders (keen to be seen getting tough on the crime their economic policies actually create) to frame criminality as a simple matter of personal responsibility requiring performatively 'tough' justice.

The social and cultural distance between those who slip into criminality and those who dispense justice is so vast that many well-meaning attempts to rehabilitate perpetrators or deter crime get lost in translation. The people with the power often possess no frame of reference for what life truly entails near the bottom, how economic conditions create cultures where crime is not simply a necessity but also morally acceptable, and few mechanisms exist to make them understand.

What's a slap on the wrist when you were beaten violently to a pulp as a child? What's losing your liberty when you live in absolute poverty and have been blackballed by every institution designed to help you? How does deterrence become operative in a mind disfigured by years of drug addiction? And how do you learn to leave a world of inebriation, violence and aggression behind when the prison rehabilitating you is just as hostile as the community you committed the crime in – and drugs are just as readily available?

Many law-abiding citizens of higher social castes assume criminality is the result of an absence of values and that people commit crime because they are bad. They believe locking them up for long periods in their teenage years with other criminals will make them less bad. Of course, such a misguided assumption is only possible if you have never viewed the world from the vantage point of an offender. Never walked alongside those who frequently find themselves within reach of the long arm of the law. This low proximity to the world of deprivation most of the prison population is drawn from is compounded by public demands for tougher sentences, which are politically easier to accommodate than calls for real action on the social and economic policies (low educational attainment, low levels of employment, cuts to youth services and community policing) which are understood to cause spikes in crime rates. In times of economic turbulence, the debate about crime and punishment becomes untethered from reality as politicians turn the discussion from root causes – government-driven poverty and inequality – and onto the easier terrain of personal responsibility.

When these harsher policies inevitably fail (because most criminal cultures arise from economic deprivation) it is then assumed it's because they still aren't tough enough, so they are dialled up once more. When they fail again, there is great resignation that troublesome youths are simply irredeemable, for how could they not respond positively? A clamour then follows for even harsher forms of retribution, with media-generated public demands for 'tougher justice', when what is required is smarter justice.

In 2018, I visited HMP Edinburgh – on paper Scotland's most dangerous prison. I met with Prison Governor David Abernethy. Like most men in senior roles, he possessed a certain swagger. One I suspect he mastered primarily because it's the kind of body language that inmates and wardens (who have more in common than you might presume) grudgingly respect. The centrepiece of his office wall was a large map of Edinburgh. Dotted all around the periphery of its forti-fied city centre and financial district were stickers denoting the areas where most of the prison's inmates arrive from – the housing estates. We walked the empty prison yard, treated to the occasional howl of abuse from the countless cells looking onto us, as he elaborated on his mission. 'Lots of the people that end up in custody come in from spe-cific parts of the city of Edinburgh; their victims come from those same places. Crime, addiction, health inequalities – they're also con-centrated in those same areas. They'll have had values instilled in them from whoever's been bringing them up and who's surrounding them. I suppose we're in the business of trying to show them that a different way can work for them.'

We re-entered the main building and took a walk through the halls, where inmates spend most of their time. The conversation turned to the theme of aspiration, which it is assumed those who end up in prison have either forgotten or have never possessed. 'I think if you asked any four-year-old in Scotland about their aspirations, every-body's got aspirations for a good life, but it kind of gets knocked out of you if you keep having failures at 5 and at 6 and at 10 and at 12 and at 16 and at 18.' David believes that aspiration still exists in these guys, somewhere. 'Our trick is to try and draw it back out again. That for me is a far more productive thing than trying to treat people like you would train an Alsatian not to do its business on the carpet,' he said. 'Rubbing somebody's nose in their past behaviours in the hope that they'll not do it again just isn't going to work, and never worked.'

David was knowledgeable, passionate and generous with his time. But something essential remained absent from his analysis. Whatever his intentions as the governor of a prison facility, his institution is but one piece of a larger criminal justice puzzle. A system which is not just terrible at communicating with itself but remains famously incapable of communicating with, or indeed interpreting, the wider society in which it exists. Policing and the courts, rather than providing deter-rence or representing rehabilitation opportunities, in fact render the custodial sentencing of traumatised young people likelier by their mere interaction with them. This speaks to the profound distance between the apparent mission of law enforcement and the courts and the dire outcomes these institutions haplessly collude to reproduce. The resistance many young criminals have to changing their attitudes and behaviour, or to taking responsibility for their part in the mayhem, is mirrored by the criminal justice system which, at its institutional core, remains a relic of the early twentieth century.

Poorer communities are overpoliced. As a result, more crime is detected and prosecuted. The social conditions associated with social deprivation also make criminal behaviour likelier. We know, for example, that crime rises in times of economic recession and that it falls in the presence of opportunity and prosperity.

Social deprivation also acts as an incubator of child neglect and abuse as well as familial adversity and dysfunction. As covered in the previous chapter, children who experience such adversity subsequently present challenging behaviours, become excluded from school and come into contact with the police – who often diagnose them as troublemakers. When this occurs, the likelihood they will come into contact with the police again and end up in the court system rises. Most young people grow out of offending behaviour but often those who do not aren't permitted to because their youthful transgressions follow them around. They are labelled, within communities, by fellow citizens, such as neighbours or local business owners, and by authorities, chiefly the police.

Susan McVie, professor of quantitative criminology at the University of Edinburgh School of Law, has been studying the notion that low aspiration is the explanation. However, there is simply no evidence to support the view that people who cannot escape poverty lack the drive or motivation for self-improvement. When I spoke to Susan, she explained: 'When we looked at children at about age 12 and we asked them about their aspirations, we found that there was very little difference between children from very different social backgrounds – they all wanted the same things.'

They desired a nice home and a car, and perhaps a holiday once a year. But as they progressed to their teenage years, at precisely the time when many of them began coming into contact with various authorities, children from less affluent backgrounds' aspiration devolved into a sense of resignation about what their future might hold. Rather than lowering their aspirations due to laziness or a lack of desire to succeed in life, their sense of possibility was constrained by a lack of opportunities, the daily grind of poverty and, in some cases, state interventions which, rather than setting them straight, compounded their difficulties. 'If the opportunities aren't there, it doesn't matter how hard you work, you're never going to manage to escape from your own circumstances,' Susan said.

McVie took part in research which tracked nearly 4,500 children in Edinburgh. Every school in the city was invited to participate – including the city's many private schools. The study found that the system itself had a negative effect on outcomes for young people who experienced a state intervention by the police, the courts, and the prison system. That once a child was flagged up as a bad apple, systems made their lives harder.

Back at HMP Edinburgh, I arrived in the arts room and was greeted by six long-term adult offenders. After the introductions, I began the workshop in the usual fashion by performing my own piece, 'Jump' – an autobiographical song. My willingness to disclose aspects of my personal experience often helps to create a safe environment for participants to do the same, should they so wish.

After my performance, we discussed writing. Each participant, in their own way, indicated that writing was 'not their thing'. That they drew blanks whenever they sat down to it. I pushed back gently against their assumptions, explaining that every time they speak, they are writing. That every time they notice something, they are writing. That all over the world, middle-class writers are tucked away in makeshift bolt-holes, desperately trying to imbue their fictional characters with traits, mannerisms and turns of phrase which come naturally to all of them. Their problem, I explained, was not that they could not write but that they had come to believe that their writing was of no value because it came from them. They were sceptical but the pens and paper beckoned, so we charged on.

One man wrote:

'This call originates from a Scottish prison,
You can hang up the phone or sit there and listen,
They will be logged, recorded and maybe monitored,

We all sit there on the other side as if we're not bothered,
But the look on your face is anything but excitement,
When you kick back in your cell and read your indictment,
I chose the wrong path,
The fast road to nowhere,
Fast cars, quick money,
Girls with blond hair,
The good stuff,
I'm a different kind of tough,
I am a diamond that has never been found in the rough.'

The room erupted in generous applause, spurring the others on. Next up was a female inmate who wrote:

'I let her down,
In here again,
And now she is gone there's no way to let her know,
That I am done,
That I am not coming back after this one,
Now I'm older and a Mother too,
I have tried not to make mistakes my Mother made,
Drink was to blame,
I always said I will never do,
But in the end the circle keeps going,
Try so hard not follow my Mum's mistakes,
But life can be shit,
All I know is to push it down,
Drown it out.'

They applauded again, though the mood was appropriately sombre. Everyone felt the weight of her regret. The heartache at a life squandered.

In many ways, workshops like these create as many problems as they solve. They address core capacities like literacy, problem-solving, social interaction and emotional intelligence, but when the participants are constantly being asked to use their own adversities, shortcomings and transgressions as a form of creative propulsion, they can remain trapped in that feedback loop, defined by what they did and what happened to them, seeing themselves as either perpetrators or victims, and nothing more. This is an example of how institutions can subtly cultivate false beliefs in the minds of those they interact with, encouraging disadvantaged people to draw inaccurate conclusions about the true nature of their predicaments.

Next came a guy who looked and sounded far older than he was. He wrote:

'I remember going to prison for the first time, I was only 4.
Stopped, stripped, searched, scratched, and clawed away by the system,
Open and closed. Doors, doors, even more doors.
Never knew how much they can bang, clang, and even slam.
Check up, sneak up, all means the same,
Time for bang up, doors closed yet again.
It's time to refrain, all this for pain.'

I found the imagery of doors fascinating – something I hadn't considered about prison, despite doors being the most prevalent visual aspect of prison life. The doors are symbolic and metaphorical. They precede and follow every meal, every visit and every journey from A to B. The door opens then is closed immediately behind you. You don't get very far in here without being confronted by one. They punctuate every sentence.

'How old are you, if you don't mind me asking?'

'I'm 31,' he replied, as I tried to conceal my shock at how weathered he appeared.

'You started when you were quite young?'

'I've been lifted since I was 5, 6, 7, 8,' he told me. 'My mum was saying one thing, I wanted to do what the scheme [housing estate] was doing.' He said, alluding to the tremendous social pressure felt by many young people in deprived communities to conform to antisocial attitudes and behaviours. Pressure which many young people successfully resist. It is often those who also seek love, approval, encouragement and a sense of security which goes unmet who are subsequently drawn into more destructive peer groups.

'What about your dad?' I asked.

'He's in here,' he said, matter-of-factly. It took a second before the magnitude of what he had said hit. Then the penny dropped.

'In here?'

'Yes,' he replied kindly, 'he's in the same hall.'

His dad was imprisoned in 1992 for an unspecified offence when he was four years old. Initially, when he claimed to have come onto the police radar at the age of five, I was sceptical but the additional knowledge of his father being sent down offers a more than sufficient explanation for his chaotic and troublesome early years. 'Last year was the longest year I spent with him in 30 years of my life,' he said.

While situations where a father and son find themselves in the same hall, in the same prison, may be uncommon, intergenerational incarceration is not. The evidence of poor outcomes for children of prisoners makes sobering reading: they are, in many cases, doomed to follow in the footsteps of their fathers – criminal justice consultancy firm Crest Advisory, in research published in 2019, found that 65 per cent of sons of prisoners end up in the criminal justice system themselves. They also estimated that there were 312,000 incidents every year of a child losing a parent to custody in England and Wales. Disturbingly, there is no formal process to identify the children of prisoners. No central record of who or indeed where these children are. Barnardo's Scotland estimates that 30,000 children face parental imprisonment every year in Scotland, with advocacy charity Families Outside noting that '7 per cent of children live through the imprisonment of a parent during their school years' and that 'there are many more children of prisoners than there are children in care.'

Care-experienced children and young people are also overrepresented in the criminal justice system. Where families do descend into cycles of adversity, involving substance misuse, alcoholism and various forms of abuse and neglect, children become likelier to be taken into care. When they are, the chance they will be drawn into the criminal justice system increases once more.

In the UK, around 37,000 children and young people enter the care system each year – 102 every day. UK charity Home for Good estimates there are around 101,500 children in the UK who are looked after away from home and that nearly 70,000 children live with almost 55,000 fostering households across the UK. The overwhelming majority of children placed in care are removed from their family due to abuse and neglect, and only a tiny number as a result of their own behaviour. While this is often done for the immediate safety of the child, being placed in the care of the local authority is extremely traumatic, not least when they have been abused by a parent or separated from siblings, and often triggers behavioural problems.

Care leavers make up 25 per cent of the homeless population and almost 25 per cent of the adult prison population. Almost half of under 21-year-olds in contact with the criminal justice system have spent time in care. Thirty-nine per cent of care leavers aged 19–21 were known not to be in education, employment or training, compared to around 13 per cent of all 19- to 21-year-olds. The risk that care-experienced children will be drawn into the criminal justice system increases not necessarily because of their own behaviour but because law enforcement discriminates against them. This is not active discrimination but occurs because police officers, like the kids they arrest and throw in the cells, are limited to a narrow range of policing responses to increasingly nuanced and wickedly complex individual circumstances, just as judges are in how these crimes are prosecuted. With little proximity to care-experienced young people, police, judges and lawyers not only draw unconsciously from a well of popular prejudice and assumption as to the root causes of a young person's behaviour, they are also compelled to act against their needs by legal frameworks and professional incentives.

Research published in the *British Journal of Criminology* found that where offending among children in care was concerned, 'the culpability of the State as corporate parent continues to emerge as pertinent.' The research conducted by Claire Fitzpatrick and Patrick Williams titled, *The neglected needs of care leavers in the criminal justice system: Practitioner's perspectives and the persistence of problem (corporate) parenting*, identified a range of personal and social problems inherent to the young adult offender population. In addition, they found that 'for those offenders with care experiences such problems were compounded by complex and acute care-related needs'.

'Firstly, the behaviour of children in care is under far greater official surveillance than the behaviour of many other children. This combines with a lack of tolerance for perceived "challenging" behaviour in some care homes, which can catapult such children unnecessarily into the criminal justice system at an early age. A report by the Howard League for Penal Reform (2016) recently highlighted the persistence of this problem, noting that those living in children's homes "are being criminalised at excessively high rates compared to all other groups of children, including those in other types of care."'

With respect to care-experienced children and young people, the convergence of overpolicing, class-profiling, and challenging behaviour rooted in trauma and dysfunction, as well as exclusion from mainstream education, effectively pushes many of the most vulnerable towards police involvement and prison. In addition, children and young people with developmental disabilities, psychological and emotional disorders, learning disabilities and additional language, speech and communication needs are well understood to be over-represented both in care and the criminal justice system. Many young people in the criminal justice system also have underlying health problems which have been overlooked as a result of systemic health inequalities. And for many, being held in custody not only uproots them residentially but makes them unemployable upon release. When you walk into a young offenders' institution you are entering a triage full of wounded children with behavioural, psychological and learning needs often so severe we have not the faintest idea what to do with them – a process which begins in a classroom and ends in a prison cell.

State interventions in early years are a predictor for incarceration later in life, yet we rarely examine the role of public institutions in this phenomenon. Instead, we study the child, dissecting their attitudes, behaviours and background. Without a parallel examination of the institutions, laws and systemic inequalities which place countless barriers in their paths, any analysis will come up short – producing more of the same.

In communities like Pilton, many youth services are effectively ploughing their scarce resources into keeping young people who struggled in school away from local law enforcement, while politicians claim 'whole system approaches' – which are supposed to involve a level of collaboration and communication between such services – are in effect. Meanwhile, in the courts, the crimes of disadvantaged youths are condemned and prosecuted year-round by judges following careful examination of all available facts, while mounting evidence that the criminal justice system fails young people on a daily basis conspicuously escapes their notice.

Not only are poorer young people overpoliced but they are also often viewed as suspects before having even committed an offence. Their mere presence elicits suspicion. Think of the stop-and-search practices which actively discriminate against young people who look or dress a certain way. Consider how police routinely disperse groups of young people for doing no more than congregating, sternly instructing them to 'move on', while knowing full well there is nowhere for them to go. From drug laws rooted in the belief you can punish people out of addiction to the low age of criminal responsibility (10 years old in England, Wales and Northern Ireland and 12 in Scotland); headline-grabbing initiatives, like threatening the parents of gang members with losing their council homes, to the systemic failure to account for how early and sustained trauma (including head injuries brought on by physical abuse and developmental complications caused by alcoholism and drug abuse in pregnancy) play a significant part in behavioural problems that leads to criminality, criminal justice systems across Britain are in dire need of rehabilitation.

Criminal justice in the UK may one day be viewed as scornfully as the churches and orphanages where child abuse and neglect were commonplace in the twentieth century. It's a racket which not only compounds but also mirrors the neglects and abuses suffered by children and young people that preceded their criminality. And it's a system filled almost exclusively with kids from working-class and poorer backgrounds.

One thing explains why childhood adversity, the care system and the intergenerational cycle of criminality and incarceration impact some more than others. One theme runs through their trauma, the formation of their early values and beliefs, their lack of security, their need to form attachments to destructive peer groups and their subsequent treatment by law enforcement and criminal justice systems. It predicts that boys, and to a lesser extent, girls, from these postcodes, are likelier to be abused and neglected, likelier to encounter authorities and likelier to descend a spiral of social exclusion as a result. If you want a clue as to what this theme is, simply sit in the public gallery of any court and contrast the accents of the people in the dock with those dispensing the justice.

The final participant, a tall man of colour, exclaims, wide-eyed: 'Poverty,' before launching into his own song, much to the delight of the group:

> 'See me when you have one foot-a shoes that-a poverty,
> No food in our belly that-a poverty,
> They take liberty through poverty.
> See we have the have and we have the have not,
> We start from the bottom and we reach for the top,
> That-a poverty.'

The rest nodded in agreement. Criminals they may well be, but they are not wrong.

4

The Law of the Landlords

The suspicion my working-class days are behind me is confirmed when the gamekeeper hands me a shotgun – it's livelier in the hands than I had expected. Stood on the rolling hills of Glen Clova in the Scottish Highlands, overlooking 8,000 acres of breath-taking wild country, I take possession of the weapon and the thought occurs that life is not so complicated after all. From this remote and lofty pedestal, amid such majestic surroundings, the world appears reassuringly simple.

'Keep the barrel of the gun in the air when you walk,' warns Mark, a third-generation gamekeeper under the employ of my hunting partner for the morning, Dee Ward, owner of the estate. 'Out here, everything is always the same,' Marks says, sage-like, as he trains his sights over the misty glens, 'so it's easy to spot when something is different.'

If only he knew just how different things are for me this fine morning, as he points off into the middle distance where he assures me a pair of golden eagles local to the area are dancing above the brow of a hill. I take his word for it as sadly my eyes are unadjusted to this natural environment – not after one instant coffee. 'Unclip it and hang it over your forearm like this,' softly spoken Dee gestures, referring to the firearm, noticing my arm beginning to tire. 'That will be more comfortable for you.'

In truth, my discomfort has little to do with the gun and more with the fact Dee is defying my expectations of how landowners should behave. I am not supposed to like him but he's a frustratingly affable chap and does not conform to any stereotype. He is gentle and chooses his words carefully. Well-spoken but not impenetrably posh. He is warm, empathic and conscientious. I am unprepared for this and his pleasant nature may yet thwart my plans to present him as the villainous foil to my righteous indignation. We don't talk party politics but from what I glean to be his central economic interests, I can only assume he is a Conservative chap. One who may occasionally be swayed by left-leaning centrist governments able to triangulate wealthier citizens by protecting the historic privileges to which they are thoroughly accustomed.

As much as I loathe inequality, blaming rich individuals for it has never made much sense to me – as cathartic as a bit of billionaire-baiting can be. Dee, and many privileged people like him, are often products of the same lottery of birth as people in Pilton. And like the youths who have come to hardwired conclusions about the nature of the society they live in, as well as their place within it, wealthier people adopt values, beliefs and aspirations in accordance with the environments and norms into which they are born and raised.

The fact we managed to secure the interview with Dee was itself a red flag that he was not the Bond villain I had hoped for; it's highly unusual that a landed gentleman voluntarily puts himself in front of a camera unless he is paying the wages of the person behind it. Dee's appearance came without preconditions and he was given no advance notice of my questions. All of this pointed to a willingness on his part to account for himself on the public stage. I guess I'd hoped that if I laid out just enough rope, I could make light work of Dee and call it a day. Instead, as we traverse the steep-sided terrain, discussing the misunderstood plight of the landed class, I am drawn into the logic of his perspective and begin to see Dee's world, however briefly, through his eyes.

We charge on. Mark falls back and out of sight. Unlike a fox hunt, the aim of the grouse shoot is not to petrify our prey by pursuing it for miles as it runs terrified for its life; a quick death for the grouse is preferable and once slain its remains are retrieved by the dogs. Grouse shooting is a sport but one which, according to estate owners, serves several practical functions in remote rural communities like this. The most obvious one being the economic activity generated. When people come to shoot grouse, they stay in local hotels and eat at local restaurants, which provides employment and career opportunities for surrounding communities. It's an argument that any environmental activist worth their salt would reflexively dismiss but given how immersed I am in Dee's perspective, and how far from the real world I currently feel, I'll buy it – for now.

Though the terrain is unfamiliar, I quickly adapt. Keen that we do not become too cosy, I pose the morning's most indelicate question, prompted by my director, who has been filming our hunt. It's the question your average punter might ask from their armchair but one which emerges awkwardly in a real-life situation. 'How much is 8,000 acres?'

'Well,' Dee sighs, like it's been a while since he thought about it, 'it's pretty much everything you can see.' The frankness of his reply is odd. No hint of a brag, even less of embarrassment. In his shoes, I'd deploy the former to cover the latter. Despite having it pointed out to me, and being able to see much of it on this clear day, the true scale of 8,000-acres is difficult to comprehend or contextualise.

This in fact marks the first time I have ever truly considered the concept of acreage. What is an acre? Where did this strange unit of measurement come from? And why haven't I been taught about it before now? I suspect I know little of acres for the same reason I know little about the stock market or fine art – people like me don't need to know about these things and so they do not feature in our 'basic' education. One acre is roughly 60 per cent of a football pitch, or 16 tennis courts in a 4x4 formation. A unit of measurement which stretches back to medieval times, acreage may baffle the average person. In the Middle Ages, an acre was the amount of land that could be prepared for crops by one man and one ox in one day. And like many of the customs and laws around property that find origins in less enlightened times, this rather crude unit of measurement (as well as the antiquated rules regarding its purchase and sale) endures to this very day.

Some of us like to think we understand roughly what acres, hedge-funds, stock-buybacks and share-liquidations mean, but our intuitions are famously unreliable. The difference between 20 acres and 8,000 acres is as hard to conceptualise as the difference between £10 billion and £100 billion – we know the latter is more than the former but little sense of how much more. This speaks to the wider apathy where the issue of landownership is concerned. If we really understood the true significance of so few people claiming so much of the natural world as their own, as well as the implications of this, perhaps we wouldn't stand for it.

I was charmed by Dee – and the grouse he gunned down was certainly tasty – but if we viewed landownership, and landowners, through the prism of personalities, we might be seduced to skim over the reports that 50 per cent of Scotland's private rural land is owned by just 432 individuals. With a population of over 5 million, this means 0.008 per cent of the Scottish population wield disproportionate power over the other 99.99 per cent. As noted by historian Jim Hunter in 2013, 'Scotland continues to be stuck with the most concentrated, most inequitable, most unreformed and most undemocratic landownership system in the entire developed world' – this much-cited quote lifted from comment he gave to the *Herald* newspaper following his resignation from the Scottish government's Land Reform Review Group.

In England, the situation isn't much better with half the land being owned by 1 per cent of the population. Indeed, for 1,000 years, the same families have owned most of England. As noted by journalist Rob Evans in a *Guardian* article published in 2019 (from which these figures are taken), 'If the land were distributed evenly across England's population, each person would have just over half an acre – an area roughly half the size of Parliament Square in central London.'

Data compiled by commercial lending director Gary Hemming and published by ABC Finance in 2019 revealed, among other things, that the top 50 UK landowners control the equivalent of Yorkshire, the East Midlands and the North East combined. The total area they own could fit Greater London 19 times over and their land accounts for 12 per cent of the UK's total acreage. If used for housing, their holdings could easily accommodate 330 million average UK homes. The land personally owned by the Queen is enough to build 1,000 Buckingham Palaces and royalty and the nobility control 3 per cent of the entire UK. The total amount of land owned by utilities companies in the top 50 list could encompass Manchester, Leeds and Birmingham twice, and the acreage owned by metals and mining company MRH Minerals alone is over 15 times that of Ben Nevis – the UK's highest mountain.

Whoever owns the land sets the price for everyone else who lives, works or plays there. And while Dee appeared every bit a man of sub-stance, not all landowners are blessed with his self-awareness – or willingness to make himself publicly accountable.

That property owners tend to remain hidden is not surprising. If not for the work of environmental activists and equality campaigners, coupled with the recent development of digital technology which has allowed for the detailed mapping of landownership in the UK, we would be none the wiser – the way many property owners seem to like it.

While the Forestry Commission and the Ministry of Defence sit in first and second place on the landowner leader-board respectively, this elite field is dominated, primarily, by individuals, dynasties and con-glomerates, which, despite their differences with respect to how they acquired or amassed their wealth, as well as what they do with it, ultimately share the same fundamental economic interests. These interests could be broadly characterised as class-based.

In 2021, as Britain continued to face the full wrath of Covid-19, journalists Rob Evans and David Pegg revealed how Her Majesty the Queen had lobbied the UK government in a bid to conceal her 'embar-rassing wealth'. Correspondence between the Palace and government discovered in the National Archives showed that Elizabeth Windsor's private legal representatives pressured elected ministers to change draft legislation to stop her shareholdings becoming a matter of public record. 'Following the Queen's intervention,' Evans and Pegg wrote in the *Guardian*, 'the government inserted a clause into the law granting itself the power to exempt companies used by "heads of state" from new transparency measures.' The arrangement, they reported, 'con-cocted in the 1970s, was used in effect to create a state-backed shell corporation which is understood to have placed a veil of secrecy over the Queen's private shareholdings and investments until at least 2011.'

Whatever the scale of her wealth – which has never been disclosed – there is quite simply no reason for a ceremonial head of state to be reviewing legislation which might affect her prior to its adoption. The Queen was also given advanced view of draft laws that related specif-ically to her personal property, her vast portfolio of private estates and 'potentially anything deemed to affect her personally'. But these revelations gained little traction, despite being proof positive that while successive UK governments have insisted that the Queen's role is ambassadorial and symbolic, she remains a powerful, unelected figure, privy to the inner workings of the British government, which colludes with her to preserve the fundamental structural inequity from which all others spring – property ownership.

A few days after meeting Dee Ward, I travelled to the town of Langholm in the Scottish Borders. Once the seat of a booming textiles industry, Langholm was a sadly familiar sight, playing host to an ageing population as young people flock to the central belt in a desperate bid to find educational and employment opportunities that no longer exist where they were born. This sleepy village is completely landlord-locked – a proverbial island inhabited by 2,400 people, surrounded by an estate owned by one of the UK's biggest and most elusive landowners, the Duke of Buccleuch.

During my visit, I met with the Langholm Initiative, a small charity with roots in the community that go back nearly 30 years. The initiative, led by environmental campaigner Kevin Cumming and Margaret Pool, was at the time engaged in an audacious bid to purchase 10,000 acres from Buccleuch Estates with a view to transforming the ailing town's economic fortunes, turning the Langholm Moor into an expansive, sustainable nature reserve. Feasibility assessments undertaken by the initiative found the reserve would lead to jobs, education opportunities and an environmental tourism boom as well as contributing greatly towards the global drive for sustainable, low-carbon growth. The community wanted to buy. Buccleuch was happy to sell. There was just one problem – a £6.4 million price tag.

On the surface, this scenario may not appear terribly problematic. People are, after all, perfectly entitled to own land. The difference between Dee and the Duke is that Buccleuch inherited every inch of his acreage while Dee took out a substantial loan at significant financial risk to acquire his. The land was bequeathed to Buccleuch by his forebears and by sheer lottery of birth, one man found himself with more power than a community of 2,400 people.

While the upkeep of such a vast estate is not cheap, government subsidies go a long way. Landowners receive all manner of handouts for planting trees, grazing cattle and re-meandering rivers. Hundreds of millions every single year are handed to landowners to provide incentives that they may develop their estates sustainably in accordance with the public good. This public investment adds more value to their private land which, when sold, profits the landowner solely. Without ascribing character or intent to either the Duke or the community buy out, the needs, aspirations and interests of both parties could be broadly described as conflicting. The town requires that the land be sensibly developed to meet the evolving economic, environmental and social needs of the people who live there. The landowner requires either millions of pounds in compensation for allowing this to happen or that no land sale occurs, thus allowing him to retain his asset. An asset that grants him disproportionate power and influence over the town, as well as guaranteeing a degree of political influence not afforded to the average person. It is vulgar and absurd. What's more, this model of land and property ownership lies at the heart of countless other inequalities and despite political mood music to the contrary, land and property inequality are rising.

Pioneering research conducted by the International Land Coalition in collaboration with Oxfam and published in November 2020 revealed the daunting scale of a global system of landownership that has allowed the top 10 per cent of the rural population to capture 60 per cent of agricultural land value, while the bottom half of the rural populations control only 3 per cent of land value. The study found that land inequality directly threatens the livelihoods of an estimated 2.5 billion people as well the world's poorest 1.4 billion people, most of whom depend largely on agriculture for their livelihoods. The report cited 'complex corporate and financial structures and cross-shareholdings' which obscured 'clear lines of responsibility for land use and management' rendering them 'harder to discern, just as they are becoming more important' and claimed that 'holding investors to account for their economic, social, and environmental impacts' is increasingly difficult when 'primary investors are unknown or geographically and institutionally remote from the land in question.' Here we see another dimension of the wider proximity problem – the accountability gap. Wealthy, powerful individuals and organisations may leverage their assets to put distance between themselves and the consequences of eschewing their social responsibilities.

While stories of community buybacks such as Langholm (which succeeded in raising funds to purchase half what they had hoped) are heart-warming and essential in granting this issue more prominence in the public mind, the scales will forever tip in favour of those who possess the vast resources to not only purchase vast swathes of land and property but to impede the political and economic reforms necessary to wrest control from their grasp, while placing themselves beyond reproach for the harms caused.

While much of the language here may relate specifically to land-ownership and land inequality there is a broader dynamic, where the interests of one group are privileged over another. This prioritisation of some over others is the definition of inequality. And from this fundamental inequity springs additional, multifaceted imbalance. Across Britain, from urban centres to rural communities, landlocked towns to far-flung islands, people with divergent interests find themselves engaged in tense negotiations across a ravine, where one holds greater power than the other: the relationship between landlords and renters; executives and the shop floor; police officers and young people; teachers and pupils; doctors and patients; counsellors and drug addicts.

Land and property ownership, rather than a peripheral issue far above the paygrade of the average punter, is central to understanding how inequalities in Britain are preserved. It is in this primary social relation that the various precedents which underscore social and economic inequality are set. And while landownership is widely regarded as a remote rural issue by many in the UK's urban centres, the institutional inequity which has allowed antiquated systems of land and property ownership to persist and expand has already had serious consequences for those of lower social classes. Not least, in recent times, for the former residents of a now infamous tower block in west London.

Emerging from the tube station on Latimer Road, my eyes wander skyward. I know what I'm looking for but haven't the faintest clue where to find it. Self-conscious that a local may catch my prying eye-line, I make for the local newsagents for some cigarettes. As I exit the shop moments later, placing a coffee on a roadside railing, I glance around again, scanning my periphery to no avail. Anxious that I may have alighted at the wrong station, I take my phone from my pocket to consult the map, turning unconsciously as I do, and am at once confronted by what I came here to see.

Rising high into the starless glare of the evening city sky, concealed in a brilliant white memorial veil, stand the remains of Grenfell Tower. For most Britons, the tragedy of Grenfell requires little explanation. Like many people, I watched the events unfold on television. The horrifying images of a blazing high-rise community left millions across the country in a state of disbelief.

Around 1am on 14 June 2017, an electrical appliance on the fourth floor of the tower block malfunctioned, starting a small, localised fire near the foot of the 24-storey structure. Built in 1972, as part of a social housing programme which aimed to clear central London, Grenfell was designed to withstand the elements. Fire-proof doors, elevator shafts and a carefully compartmentalised interior meant any fire that did break out would, in theory, be contained to whichever part of the building it started. There it would be allowed to burn as the alarm was raised and residents were evacuated. But on this morning, that did not happen. Onlookers watched helplessly from the street as their neighbours, unable to escape their flats due to the intense heat of the flames and an acrid, thick wall of smoke, appeared intermittently at windows around the building, calling for help before retreating from view.

Text messages and phone calls were made to friends and family by many in the building as they began to confront the reality that they would soon perish. As the intensity of the fire increased, others saw only one way out and in either their final bid for survival or simply their hope for a less painful, undignified death, leapt from their windows.

Think for a moment what this must have been like. Close your eyes and imagine how you would console your children. Try to visualise the very home you live in burning down around you and those you love, where the only certainty is that help is not coming. At this precise moment, when many Grenfell residents were faced with the brutal and wicked choice of leaping to their deaths simply to escape a more excruciating demise, most Britons lay oblivious, asleep in their beds – just as the victims had when the fire began its deadly spread.

An ominous pillar of thick, grey smoke bellowed silently into the night sky as pockets of traumatised residents fled the building from the lower floors, many passed by firefighters pouring in to ascend the dark, solitary, smoke-filled stairway, in a courageous bid to locate and evacuate survivors. News of the blaze broke just before dawn, at around 4am – three hours after the fire had started. As the first live images were broadcast, showing firefighters on raised hydraulic platforms training their hoses on the only section of the building which hadn't gone up, it was already too late. With just one staircase and two-elevators serving around 600 people in 120 flats, the loss of life in the event of such a catastrophic event was a foregone conclusion. The highest fire-service ladders could barely scale half the building.

Not even during the Blitz, when London was brutally bombarded by the Nazi war-machine night after night, was a building ravaged by fire in this way. Yet somehow, in Lancaster West Estate, a fire that started on the fourth floor engulfed a 23-storey building in 20 minutes. The fire itself may never have been preventable but the loss of life most certainly was. Local people here know it. After all, they had tried desperately for years to raise the alarm on the tower's dire fire-safety issues but were ignored. They published an article later in the morning, as images of the tower swept the globe, titled 'Grenfell Tower Fire', in which they claimed to have warned the local authority and the private company responsible for the safety of the tower block, on numerous occasions, that fire safety was a potentially lethal issue:

> 'Regular readers of this blog will know that we have posted numerous warnings in recent years about the very poor fire safety standards at Grenfell Tower and elsewhere in RBKC. ALL OUR WARNINGS FELL ON DEAF EARS and we predicted that a catastrophe like this was inevitable and just a matter of time.'

Below it, they posted links to no less than ten previous blogs, dated between 2013 and 2016, in which they attempted to warn the Royal Borough of Kensington and Chelsea (owners of the tower) and the Kensington and Chelsea Tenant Management Organisation, who ran it on their behalf. Prior to the fire, the Grenfell Action Group had been engaged in a wide-ranging campaign spanning years, which aimed to draw the attention of the local authority to the concerns and frustrations of people on the Lancaster West Estate. Their attempts to bring attention to unglamorous issues like air pollution, the closure of the local public library and, most presciently, fears that the contested refurbishment of the tower was mired in a culture of corporate negligence and corruption that presented a danger to human life remain there for all to see. Some of you may find this hard to believe but those of a certain social class won't be the least bit surprised.

On this chilly February evening, Latimer Road is busy. Nobody else seems to notice the tower. As I sneak another glance, anxious I may look like a disaster tourist, I have a self-important sense that I am bearing witness to a spectre nobody else can see. The sight of the tower is transfixing, temporarily muting my other senses. Grenfell, for a time, is all I perceive.

Curious as to how I might get closer to the foot of the building, I scan the local surroundings. What becomes immediately apparent – despite Grenfell being within minutes of my position – is that the structure is encircled by layer upon layer of urban obstructions. These obstructions are not, as you may think, the result of recovery work in the aftermath of the fire but are in fact part of the community's haphazard and improvised design.

Until the fire, Lancaster West Estate had been undergoing a contested process of regeneration, much of it taking place around or on the tower itself. Many local people suspected the work – which included the installation of insulation on the building's exterior – was being carried out not for the benefit of the community but for the more affluent residents on its periphery, who might gaze upon it from a distance. The estate is nestled within an area of London more noted for its fashionable boutiques, pricey florists, cafés and museums. The refurbishment was just one of many sensitive issues in play at the time of the fire; much of the regeneration around Latimer Road – including the school and leisure centre – was unpopular among many locals. It involved the temporary closure of pre-existing facilities that many were satisfied with or relied on.

What is immediately striking is how close to the foot of the tower much of this redevelopment took place. You get an immediate sense that accessing Grenfell might be tricky for anyone unfamiliar with the area. More so, if you happen to be at the wheel of a fire engine despatched from the other side of London, racing towards the deadliest inferno in recent British history.

My guide arrives and we walk east towards the school where a crowd is gathered for a Silent Walk of remembrance. This is a regular occurrence in the area since the fire. As well as paying respects to those who lost their lives, the walk is a demonstration of community solidarity and resolve. It's easy to pull together when the whole country is behind you. The real test of a community's character comes when the news agenda shifts and its plight quietly slips out of the public mind. Aside from the ongoing public inquiry, Grenfell last made headlines in November 2019, when a video began circulating online appearing to show a group of men celebrating Guy Fawkes night by burning an effigy of the tower. During the disturbing clip, a person can be heard saying: 'Didn't it start from the tenth floor, though?' while others chip in, crowing: 'Help me! Help me!' and 'Jump out the window!' As the bonfire engulfs the effigy – complete with models of terrified victims – a man says: 'All the little ninjas getting it at the minute,' before adding, 'That's what they get for not paying their rent.'

The video spoke to the racial dimension often overlooked when the story of Grenfell is retold, and how it intersected with class. Grenfell was originally built as council housing and the majority of residents at the time of the fire were working class or poor. Eighty-five per cent of residents who perished were from ethnic minority backgrounds, which is unsurprising when 40 per cent of high-rise residents in the social housing sector are from BAME communities – despite making up just 14 per cent of the population. The disaster disproportionately affected ethnic minorities who were, by virtue of their racial identity and social class, likelier to be living in poor-quality housing at the time. Race was believed to be such a significant factor that Imran Khan QC – representing the bereaved, survivors and relatives when the scope of public inquiry into the fire was first being discussed – urged the home secretary to broaden the enquiry's terms of reference to examine whether race, religion and social class played any part in the events surrounding the fire.

While many on social media were quick to dismiss this concern, there is a historical precedent for such a request. The murder of Stephen Lawrence, prior to the public inquiry held years after his death, was widely regarded as not being racially motivated. However, the Macpherson inquiry into both the murder and the criminal proceedings that followed found that race played a significant factor in both Lawrence's death and the subsequent criminal investigation. Macpherson concluded that the Metropolitan police force was institutionally racist. The case of Stephen Lawrence cast new light on the racial discrimination embedded in British culture, expressed in our laws and institutions. Discriminations which are compounded by class.

As we prepare to set off on the 60-minute pilgrimage from Latimer Road to a corner of the community which now acts as both a shrine to victims and an unofficial public square, I eavesdrop on the conversations going on around me between local people, many of whom lived in Grenfell or lost loved ones. A theme emerges from whatever small-talk I can glean: the council. It's clear that while politicians have once again seized the reins of the narrative, Lancaster West Estate has become a battleground where local people and the council are engaged in a cold war of sorts.

One proxy for the ongoing conflict is the local authority's roadside signage, which is regularly defaced before an outsourced lacky is deployed to parachute in by nightfall to repaint. This civil disobedience is connected to the broader issue of public space, which is hotly contested in the estate. In the immediate aftermath of the fire, the community utilised many of the walls in the area to hang missing person posters, witness testimony and tributes to the dead. Having reclaimed many of the concrete surfaces, some locals conducted their own investigations into the blaze, using the walls as notice boards, detailing the decision-making processes and chains of accountability between the complex web of private construction companies, property developers and council officials they believe were responsible for the catalogue of terrible decisions which preceded the tragedy. They contained names and dates, as well as purported evidence of corruption. The council removed them immediately.

Silence falls as the gathering crowd swells to around 400 people and we set off in unison, bringing the early evening traffic and pedestrians to a respectful standstill. The urban surroundings begin to shift as we cross the territorial lines which subtly segregate communities by social caste. We leave the London depicted in *EastEnders* and *Only Fools and Horses* and in a few blinks of an eye emerge in what feels like the set of a *Paddington* film. It is here, at these road junctions and railway stations, where the realms of rich and poor diverge so dramatically. Millionaires and food-bank users, comfortable home-owners and struggling renters, the old whose mortgages are almost paid off and the young who can't get a mortgage because the banks won't lend – here, at these historic socioeconomic intersections, they live side-by-side, in striking physical proximity yet separated by gaps in wealth and life expectancy so stark, they make the shocking disparity in house prices seem almost comical. In the plusher postcodes, a boy can expect to live until he's 91, but one born on the other side of the tracks in a poorer area may only see 76. A flat on one street may be worth £125,000, while just a few moments down the road a similar property can go for £2 million.

Property and the cost of housing are absolutely central to the maintenance of inequality in Britain and key to understanding the forces which made the Grenfell fire likelier. Property is about more than simply owning a home; it's about market forces which cleave divides in the population, placing them at odds with each other socially, economically and culturally. Whether you own one home or you own five, your assets grant you a distinct advantage and one which is contingent on the inability of others to acquire one.

The problem today is that the group of people who possess the means to purchase property is shrinking, while the group forced to rent is growing. If you own a property and rent it out, you draw a passive income from your tenant. If you have a decent job, this gives you a dual income while your tenant must work to pay more than half of their sole income simply leasing your property from you. In the past, renting was seen as a short-term option for many on the way to buying their own home, but in times of economic uncertainty, banks demand larger deposits, which increasing numbers can no longer save for due to the increasing cost of living. The problem is compounded by the fact tenants cannot build a credit rating by reliably paying their rents and utilities as agreed, creating absurd scenarios where mortgages that often cost less than their monthly rents are denied. People become trapped in handing over their incomes to the property-owning classes, one of many mechanisms by which wealth is transferred upstream. While landlords come in all shapes and sizes, and implying they are all rich or that their sole motivation is ruthless profit extraction is unfair and inaccurate, even the millions of well-meaning property owners, who aspire to treat their tenants fairly and whose only aspiration is to pass something onto their children, will be corralled by the norms and moral hazards of the housing market into practices which contribute directly and indirectly to the wider housing crisis.

The cult of homeownership is extremely exclusive. For decades, homeownership has been the preferred model of governments, which have leveraged their institutional muscle to incentivise it, in schemes such as Right to Buy, Help to Buy and state partnerships with banks and building societies to create ISA schemes which reward saving for a mortgage, as well as quirks in the tax code over the years which aimed to encourage homeownership. As a result, social housing is now regarded as inferior, culturally. People don't just want to own their own home because it makes economic sense but also to project a symbol of status and success. Great prestige is therefore conferred to those areas where house prices are high but one factor which determines whether house prices rise or fall is the overall housing supply, which has, for decades, been outstripped by demand. Properties increase in value as a direct result of social and affordable housing stock being neglected. It therefore becomes politically counterintuitive for governments, local or national, to address the economic roots of housing and property inequality as it would involve action that may disturb the lifestyles of the most politically lucrative slice of the UK population – property owners.

In the case of Grenfell Tower, the disparity between many of the residents who rented there and the wealthy middle- and upper-class property owners who populate the surrounding areas lies not merely in obvious metrics such as house prices, wages or life expectancy but also, crucially, in political representation – the interests of those who own property consistently take precedence over those who do not. This is reflected in the various perks enjoyed by property owners, who politicians are ever keen to keep on side. During the pandemic, property owners were well looked after while activists had to fight for an eviction ban to prevent landlords from turfing people out on the streets. In Lancaster West Estate, this unpalatable truth around political exclusion lies at the root of the tragedy – everyone has a voice but the more property you own, the bigger your megaphone and the keener politicians are to hear what you have to say.

The walk ends at the memorial wall beneath a rail bridge, another of the contested areas which the community has occupied. The names of the victims are read aloud, the majority of which are working-class people of colour or refugees. 'I saw the flashing lights outside my window,' an eighth-floor resident has written on the wall. 'I saw smoke, I saw sparks and flames up on the sixth to the seventh floor. I looked out of my door to see if anyone else had woken up – that's when I saw no one else was woken up. Everyone was asleep.'

As a small army of journalists descended on the estate in the early hours hoping to make sense of the tragedy, many of the heartbroken, grief-stricken residents already had their suspicions about the fire's rapid spread through Grenfell. The fire was, in fact, confirmation of their worst fears. This was also true of many working within the housing sector, who had long campaigned for a re-think of both refurbishment practices and fire safety generally. Peter Apps is deputy editor of *Inside Housing*, a trade magazine published weekly for professionals in the housing sector. He is frank in his assertion that the media overlooked the fears of local people because risk is not deemed newsworthy.

Grenfell is often presented as an isolated, freakish event. There is relentless focus on the fire's rapid spread and implicit in this is the assumption that it could not have been foreseen, which is patently false. The truth is, the clues were everywhere but nobody with the authority to act, nor the influence to leverage those responsible for the building – the local authority – saw fit to take action until the bodies were being removed.

The Grenfell Tower blaze claimed the lives of 72 people and brought the community to the cliff-edge of civil unrest. Only then was the issue of cladding, central to the fire, deemed to be newsworthy. Apps had pitched articles to both the London *Evening Standard* and the

BBC in the weeks before the fire. 'We got hold of documents showing a fire at Shepherd's Court was linked to a flammable panels attached to the outside,' he says, 'and the London Fire Brigade was warning other councils in London to check their stock. Sent to BBC London as well – neither replied.' Of the *Evening Standard*, Apps adds: 'They actually did a front page on it in the end, but not until after Grenfell.'

As the flames were beaten back by some 200 firefighters, despatched in 40 units from across London, and the fire was finally brought under control, the human cost of the disaster played out on breakfast television in the haunting shadow of the smouldering edifice. The community of Lancaster West Estate immediately became synonymous with the doomed tower block that brought it to wider public note, though most people then (and to this day) couldn't tell you the name of the community where this tragedy occurred. For many, west London is as specific as it gets. Since the fire, most of us (if we are honest) know and refer to this entire community as 'Grenfell', despite that name applying only to the tower block itself, built in 1972. The extent to which the voices of this community had been routinely ignored played a key role in the sequence of decisions that led to the fire, not least the choice, made in the name of cost-saving, of flammable cladding and insulation materials that encouraged the fire's rapid, deadly spread.

Many online were quick to warn against 'politicising' the blaze and dismissed with alacrity local people's 'paranoid' assertions that the cladding was fitted to give the neighbouring community something less hideous to look at. The notion that gentrification (a practice which aims to raise house prices) was a factor in this disaster was laughed off by the same people who disputed that Grenfell was a deprived community – because people who own IKEA furniture can't be poor. A quick glance at the 2014 planning application, confirms the 'paranoid' locals were, again, correct: *'The changes to the existing tower will improve its appearance especially when viewed from the surrounding area.'* While gentrification was not a central factor, and it is customary for wider benefits to be noted in any planning applications, it's clear that this created an additional sense of resentment for many locals. It is also notable that the fire-safety risks for residents which had by then been well established by insurers, were not featured in the document but the aesthetic benefits for wealthier homeowners in the vicinity were.

The morning of the fire, as a small but very loud countermovement of armchair construction and fire-safety experts warned against a rush to judgement, the cheap cladding widely regarded as the cause of the fire's rapid spread rained down from the tower, onto traumatised, grief-stricken families, friends and neighbours like hellish confetti. People were picking the polystyrene out of their hair.

On my way back to the tube station, we take the long way round. Finally, we reach the foot of the tower where flowers and condolences remain in pristine condition. I look up at Grenfell hoping for a revelation of some sort but instead my mind intrusively projects the horrifying television images seared in my memory onto the surrounding landscape. I hear the children screaming. I see the bodies falling. I feel the collective heartache and disbelief of a community that was left for dead. And I feel their anger. Their fury. And their desire for retribution, in the form of accountability. A local woman approaches and hands me a tea candle, saying nothing, before returning to her flowers. There is something unsettling about commemorating this tragedy with a naked flame. My guide then draws my attention to a poem titled 'Life in Monochrome', written anonymously and taped to the wooden barricade in the foreground of the tower. Over the wooden partition Grenfell looms, as my guide leaves me to read it alone.

The political and economic exclusion of working-class people and ethnic minorities lies at the root of this historic tragedy. Exclusion which occurred because local government was too attuned to the interests and aspirations of wealthier citizens and private enterprises, not simply in relation to Grenfell Tower, but with respect to every public asset in the area.

In 2021, the Grenfell Inquiry heard that the council, prior to the fire, and in an effort to increase revenues, had announced plans to sell a number of leases on vital public assets – on which local working-class people depended – to private school operators, including Notting Hill preparatory school, which charges £21,000 a year per pupil. Also put up for sale was a public library, a teacher training centre and space in a Citizens Advice bureau. Some of the plans were dropped in the aftermath of the fire and the council has since apologised, admitting that 'too often the council put the narrow goal of generating commercial income above the broader aim of delivering benefits to our wider community', that it, 'fell below the bar on consultation, transparency, scrutiny and policy' and that it could not say, 'hand on heart that residents were involved every step of the way, or that the council put their interests first and foremost, and for that we apologise'.

Consider the fact that so many private enterprises, from independent schools to property developers, would so quickly encroach on the receding public realm to extract profit and expand their fiefdoms, enabled at every step by the local authority.

Meanwhile, as the tower smouldered in the skyline, and hundreds of working-class people became homeless, thousands of private properties surrounding them lay empty. One woman even wrote to Foxtons, a major estate agent in the area, in the immediate aftermath of the fire. 'They had enough local empty properties on their books to house all the people in need immediately,' she told me. 'I am still perhaps naive or perhaps hopeful enough of the goodness in human beings that I thought this idea might work,' she added, 'that localised generosity might prevail amongst people who owned empty homes in the local area to help people from Grenfell. But no. They didn't even reply.'

The disproportionate power of those who own property and those who rent was perhaps best exemplified by how quickly calls to temporarily seize (reacquisition) and repurpose the countless vacant properties in the area as accommodation for those made homeless by the tragedy were dismissed. We are talking here not simply about properties which were temporarily empty, soon to be rented or sold, but about houses purchased not for living in but for the sole purpose of wealth accumulation. Stop and think about what this really means: requisitioning property (a practice which occurs during times of national crisis, like war) to temporarily house the victims of a disaster represented an egregious infringement on the rights of property owners not simply to own an unlimited number of properties but to purchase them with no intention of letting anybody live in them. A property owner, therefore, has more right to buy up homes and use them as financial safety-deposit boxes than someone recently made homeless through no fault of their own has to be housed.

This practice is not only a moral perversion but an economic one; the artificial inflation of house prices in the surrounding areas, and the knock-on effects this has on the cost of living, contributes massively to entrenching the vast income, educational, housing and health gulfs we see between the wealthy and the poor (and the gaps in political representation) which are central to the Grenfell tragedy. When your fate lies in the hands of a landlord, whether they own 10,000 acres of wild country or 20 empty flats in a London borough, and no incentive exists beyond profit-motive, you are effectively living under a form of serfdom – dressed up in the self-regarding jargon of free-market economics. The people of Langholm and the people of Grenfell, despite existing in worlds which on the surface could not be more different, faced the same fundamental conundrum: their rights to their lives and livelihoods came secondary to the rights of their landlords to do as they pleased.

In the Grenfell Tower fire, we see the hellish convergence of myriad inequalities in housing, property, race, class and political representation. We see the experiential gap between those who regarded the tragedy as an unfortunate series of mishaps that could have happened anywhere and the survivors and their families who know it happened to them because they were poor.

But in the end, it's the accountability gap which may prove to be the hardest one for the community to accept. Years later, the 'paranoid' locals' concerns about gentrification, the stripping and selling off of vital public assets, and the repeated political exclusion which underscored all of it stand vindicated. Meanwhile, many of the same forces, who contributed significantly to the cultures of corporate neglect and political corruption which preceded the fire, continue shamefully in their collusion to retrofit the horrifying truth of the Grenfell Tower tragedy with a pathetically fashioned, face-saving exterior. One wonders how many of them will ever see the inside of a prison cell.

5

The Shame of Homelessness

'No one is to go begging in the state. Anyone who attempts to
do so, and scrounges a living by never-ending importunities,
must be expelled from the market by the market-wardens, and
from the surrounding country, conducted out by the country-
wardens across the border, so that the land may rid itself
completely of such a creature.'

Plato, The Laws

'Good morning, Tracy, how are you?' I ask, as a young single mother
opens the door of her Edinburgh flat.

'Not too bad,' she replies, her breath visible in the air. It's a chilly
winter morning but all hope of some respite from the cold is dashed
upon entry – Tracy's home is freezing. Stood in the middle of a bare
living room watching television is Zak, her one-year-old son. He is
not a happy bunny.

'It must just be miserable for him being that cold,' I say.

'Yeah,' says Tracy, 'it's colder in here than it is out there.' I decide to
keep the hat and gloves on as I'm given a quick tour of their temporary
accommodation where they have been placed by the local authority
after becoming homeless. 'The heaters are terrible,' says Tracy. 'I can't
have him in this house, he's going to catch pneumonia.'

The national homelessness charity Shelter Scotland found that, in
Scotland, households like Tracy's are made homeless every 18 minutes, and of those placed in temporary accommodation, over 6,000
are children. She says heating her flat costs around £9 a day, or £280
a month – nearly half of the money she receives on benefits. 'He's been

riddled with infections. Ear infections, throat infections, airways infections, bronchitis,' Tracy looks and sounds exhausted but is in good humour. Then again, you'd have to be, or you'd go mad.

She shows me Zak's room. 'They're telling me it's not damp,' she explains with a look of utter bemusement as she draws my attention to the visible condensation gathered in a puddle on the windowsill. The house is poorly furnished and feels scabby and unhomely as a result. Tracy and Zak have few personal belongings here. We return to the living room where the presence of the television gives the illusion of warmth. Tracy produces a piece of paper. 'This is Zak's letter from the doctor saying that he's had respiratory infections due to the dampness in the house.' The damp that Tracy is being told does not exist. The letter is sternly written and unequivocal. Tracy's one-year-old son's health is plagued by the respiratory effects of the cold, damp conditions. 'And I've told them I've got this letter and they're like,' she shrugs, 'don't need moved.'

Due to the chronic lack of social housing, in the last five years the Scottish government has spent over half a billion pounds of taxpayers' money placing those at risk of homelessness in temporary accommodation like this. That money does not go to the tenants but directly into the pocket of the private landlord who owns it – Tracy's landlord is charging almost twice the cost of social housing. 'I don't understand why the government are wanting to pay money for people like me to stay in a house like this. My housing benefit is paying £850 a month for this when they could put me in a council house and save themselves half the money. Its only £450 for a council house.'

The problem is, of course, that Thatcher's Right to Buy scheme, which allowed council tenants to purchase their properties at eye-watering discounts, was not followed by a strategy to replenish social housing stock and has therefore led to a chronic shortage of housing – creating a two-tier system where owning your home has become the ideal and where social housing is regarded as the preserve of the poor. In Scotland, the Right to Buy was only recently repealed by the Scottish government in recognition that too much social housing stock had been lost, driving up rents and causing residential instability and homelessness.

Tracy and Zak arrived here after a relationship breakdown with a former partner. But as is increasingly the case, her mere interaction with officialdom, in this case Edinburgh City Council, has placed her son at risk of health problems and various forms of stress which are sure to impact on his early years' development. Zak has not even reached pre-school age and the myriad disadvantages he faces are palpable. Tracy, too, is struggling with her own mental health issues. Little disrupts a mother's wellbeing like the suffering of her child. This is coupled with the recent breakup which made her homeless and the unending stress of dealing with government bureaucracies, complicated in Scotland where matters like housing are devolved but some areas of welfare remain reserved.

'What's your hopes for the wee man here?' I ask.

'To be honest with you, if I can get a house then I'll be happy, I'll be able to get back to work, get him organised and not be cold all the time, not well.'

As Tracy speaks, I feel emotionally pained by the sheer simplicity of what she requires yet is denied. All she wants is a stable, modest home so that she can return to work and provide for her child. Tracy understands that her son's early years are crucial. 'I'm hoping I can get out of it before it makes a big impact on him,' she says. 'Hopefully, I can be out of this situation before he's in nursery and things like that so he's in a warm house.'

The practice of local authorities renting substandard homes from landlords and temporarily housing vulnerable people in them due to a lack of social housing is widespread across the UK. When we have a system where a single mother and her young child are being moved into temporary accommodation that costs £850 a month, and for that they are getting dampness and freezing temperatures that are affecting their health, we must ask: what on Earth has gone wrong? Tracy's benefit money is effectively being laundered through her and then distributed to various private enterprises which are, due to the dysfunction of the free market, charging well over the odds for the goods and services they provide.

Utilities companies and private landlords leave Tracy with very little to live on. But this, unfortunately, is no accident. This kind of society, where the free market dictates everything and the state exists solely to mop up however much of the mess it can, has been the policy of successive governments across the UK for decades. This naive belief that markets will always self-correct without state interference is why so much UK housing stock was sold off. It's why private developers were unleashed to build 'affordable housing' – much of which is not affordable – and pre-existing social housing stock was neglected. It's why successive governments across the UK have invested billions creating incentives for people to get on the housing ladder while cutting billions from the welfare budget that supports poorer families with living costs.

Now governments are playing catch up, in recognition that modern, safe and affordable housing is essential, not simply because it makes economic sense but because the chronic shortage in social housing is a contributing factor in the wider homelessness crisis.

It's the winter of 2018 in the north-east of Scotland and we're out filming – because we're gluttons for punishment. Aberdeen, also known as the 'Granite City' (thanks to the distinctive colour of its buildings, constructed using locally quarried, grey granite), is a city of contradictions and extremes. Its traditional industries of fishing, textiles, shipbuilding and papermaking made it a top seaport city throughout the industrial period. Traces of the harbour's working-class history can be found in the weathered waterfront bars which still attract a decidedly battered clientele, but these days the everyman has no need to be near the harbour – unless in the business of drinking. Hulking, brightly coloured ships, so tall they obstruct the view of the sea, move in and out day and night, but the cold north-easterly waters conceal a bounty far more valuable than fish.

Rich in valuable fossil fuels, Aberdeen was crowned the oil capital of Europe in the 1970s and, as well as oil and gas, is widely regarded as a pioneer in the field of sub-sea engineering. Set on the gusty east coast, looking onto a choppy North Sea, the horizon-line adjoins the waves and the heavens, vanishing in a rapturous wall of blue whenever the sun is shining. On a bright day, beneath an immaculate, cloudless sky, Aberdeen may be Scotland's most beautiful city; sunbeams reflect off the granite, rendering even the most ordinary building prestigious and majestic. This is no such day. The sun can barely work up a heat and the darkening clouds, which only accentuate the monolithic grey, are threatening rain.

Not far from the coastline, tucked away in a congested, chaotic industrial district, staff and volunteers begin loading up a van. 'The milk usually only has a couple of days before it goes off,' says staff member Graham, as he chucks an insulated container unceremoniously into the back of the vehicle, 'so we need to get it out to people as soon as possible.' The Community Food Initiative operates beside the harbour, distributing donations to various other organisations throughout the day. Graham is assisted by a young man, Danny, who was recently compelled by the court to carry out a community disposal order at the food bank. 'This is redistributed food from the dairy industry so if you can grab a box and we'll get loaded into the van,' says Graham. We load in ten or so boxes before taking our seats up front. 'We're heading to a wet hostel,' I'm told – a supported accommodation that works primarily with people experiencing alcohol problems, at risk of becoming homeless.

During the short drive, both gentlemen give me their respective takes on the issue of food poverty. 'So, we published a report last year,' Graham explains, 'which points to the implementation and administration of welfare reform as the key driver for the increase in food-bank use. And delays in benefit is a big one, now with Universal Credit, which has been rolled out full service in Aberdeen now, we anticipate food-bank use to increase further.'

Graham is painfully restrained, indicative of most who are employed by the third sector, where speaking truth to power can quickly earn you pariah status. Nonetheless, he correctly identifies welfare reform as central to the current wave of food poverty. Oil-rich Aberdeen has more food banks than any other Scottish city.

As Graham explained, the biggest reasons for food-bank use are low income, delays in benefits and changes to benefits. Compounding this is the confluence of wage stagnation and unaffordable housing set against the rising cost of living. Four years ago, Aberdonians earned the highest salaries in Scotland, coming second only to London at UK level in terms of the average pay packet. Today, things are rather different. Welfare reform, wage stagnation and a housing crisis have since coalesced, creating a tidal wave of adversity. The fact Aberdeen remains the most expensive Scottish city to live in doesn't help. From a peak in 2014, the subsequent collapse in the price of oil left

Aberdeen with one of the fastest rising rates of unemployment any-where in the UK. Since the 1970s, oil extracted off the coast of Aberdeen has generated hundreds of billions in revenue – much of it yet to trickle down.

As we approach the hostel, Graham's actions do the talking. He is effectively playing a dual role as a charity worker and mentor to Danny – a young man struggling to find work because of a criminal conviction. He is more candid than Graham, having yet to learn what it wouldn't do to say if you desire a career. I ask Danny how many hours he got from the court. 'Three hundred,' he says, and he is no closer to finding a job. 'I find it quite difficult to be honest. Scared to go into employment. "Have you got a criminal convictions?", well you don't want to lie, do you, and you say yeah . . . I think everybody deserves a chance even if you've got a criminal record, even if you don't have any qualifications, people deserve a chance. There should be a lot more support and a lot more help than there actually is.'

Danny's understanding of the issues is more visceral and urgent. What he lacks in a coherent socio-political analysis he makes up for with self-awareness and lived experience. Without realising it, the two men dovetail one another effortlessly, creating a fuller picture of how the current landscape has been shaped and, furthermore, how it is being experienced by those trying to navigate it.

We arrive at the supported accommodation service and quickly unload the van. A chef meets us at the door, directing me in with one of the boxes, which I leave on a table near his kitchen. As he transfers the milk from the container to fridge, he tells me the service, also run by a charity, is currently supporting 15 'clients' – people with alcohol and drug use problems who are experiencing residential instability as a result. The laws of supply and demand which have transformed Aberdeen's industrial fortunes over the years have also led to an abun-dance of alcohol sale points, offering high-strength products at bargain-basement prices and illicit, often deadly, drugs, as those unable to get an economic foothold increasingly turn to oblivion to escape the shame of their ever-visible social exclusion.

We say our goodbyes and return to the office, which feels more like a newsroom; around 20 staff work frantically though in good spirits, getting through the business of the day, sustained by coffee, biscuits and the chatter of local radio. The term 'food bank', like much of the common parlance used to describe British social problems, does not convey adequately the true scale of the problem. Food banks are more like small factories, employing staff while relying on volunteers and members of the community to manage a massive daily operation that involves collection and distribution of food and other goods, as well as accounting, stock-taking, fundraising and advertising. In truth, handing out food is probably the simplest aspect of what a food bank does.

But like many charities and volunteers, they are fighting a losing battle. A quarter of all referrals to this food bank are as a direct result of Universal Credit, according to volunteer Kerry Wright. Specifically, she pinpoints the digitisation of the benefits system as a key factor. 'Everything is being digitalised now in the benefits system. There's a lot of people out there who don't have access to a computer, don't have access to Wi-Fi, so they're just being cut off.' She's seen countless people whom, if not for third sector organisations providing an access point to apply, would fall through the net of social security altogether. And this move to digitisation reveals perhaps the greatest absurdity of austerity Britain – you cannot own a phone if you're poor but you can't access benefits without the internet.

What becomes abundantly clear is that staff and volunteers across the city are battening down the hatches and redeploying whatever resources they have available in a desperate attempt to absorb the social devastation created by UK government cuts to welfare and public services. Savage cuts which the Scottish government, despite its progressive rhetoric, is all too keen to hold its nose and pass on. The most disruptive and heavy-handed benefit change in the history of the Welfare State is now the central preoccupation of most charitable organisations. This at precisely the time when many organisations are facing closure.

One such service is Margaret House, a supported accommodation project run by the council. I arrive to meet John, one of 12 tenants with a room in the service, which will close its doors for the final time early next year. 'I do try and keep my room tidy,' John says, opening the door to his small but neat room, 'we actually have an inspection once a week.'

'What was your life like before you became homeless?' I ask.

'I was in full-time employment,' he says. 'I worked in mental health nursing looking after elderly and younger disabled general nursing. That's what I did for 25 years of my life but that seems to have slipped away from me.'

You'd never assume John was once a mental health worker – homeless people and those with addiction problems are viewed only through the lens of their afflictions, rendering them less relatable to the average punter, placing additional distance between the most vulnerable and a general public, which often misunderstands the nature of their problems. John is managing alcoholism and homelessness which are both compounded by the social stigma which arises in this proximity gap. 'People forget that before we landed up in that situation, we have lives that we lived, we've got jobs, we've got family.'

John has five children, three girls and two boys. 'They all live in Aberdeen,' he tells me, with a glint of shame in his eye. I ask if he's seen them recently. 'I always keep my eyes open that my children aren't coming down the street because I do . . .' he pauses, choosing his words carefully. 'I would find it embarrassing,' he continues, 'if they saw me sitting begging, I would find that hurtful.'

We leave the service and toddle up to Union Street to do a spot of begging. 'What I normally do is go to this skip,' John tells me, a slight spring in his step that someone has taken an interest in his occupation. He walks directly to the two wheelie-bins parked in the lane behind Starbucks.

'What are you looking for?' I ask.

'I'm looking for a begging cup,' he tells me, opening the first bin and lifting one out. 'And I'm also looking for a bit of cardboard,' which he locates in the second bin. I am taken aback by how methodically he approaches begging. There is a craft. A right way of doing it, which increases the success rate.

I sit on the pavement beside him, to get a sense of Aberdeen's prestigious commercial centre from the eyes of one of its most marginalised citizens. It's cold and bleak. People cannot help but look down on you as they walk past, trying not to catch your eye, fearing they may be drawn into an interaction. It's not long before an acquaintance of John's walks by. They say hello and get to chatting before John sends him into the local shop for a bottle of wine. Without alcohol, John experiences mental and physical withdrawal symptoms, meaning much of the proceeds from begging are used to purchase booze – not to get him drunk but to keep him alive. That his friend would willingly purchase him alcohol, despite understanding he is a chronic alcoholic, may appear from a distance a thoughtless and even harmful gesture, but in this world of marginalisation, such a deed can be an act of profound solidarity and even love.

What fascinates me most is that John persists with begging and even rough sleeping despite having access to safe, dry and warm accommodation. Why doesn't he remain there where he will be safe? From a distance, people may regard this self-defeating behaviour as a lack of willingness or a bad attitude. However, when you take into consideration John's current needs, his behaviour is deeply logical. There is a curfew at the service but John's alcoholism and addiction issues require that he obtains a drink or a fix every hour or so throughout the course of a day or he risks entering withdrawal.

It's in the warm exchange between John and his acquaintance that the main reason he has not yet settled in St Margaret's House becomes evident – John is drawn from the seeming safety of a homeless service and back onto the streets by a community who share his experiences, who are often better placed to meet his immediate needs. Marginalised people operate on the very periphery of society and in their exclusion, they become bonded, like the Pilton boys. They face the same daily dilemmas and use the same language to describe them. A solidarity develops, rooted in their shared knowledge that few will ever truly understand what it's like to be them. John is drawn away from safety and into danger not simply by addiction, but by the pull of the few social connections he has left.

As the suns sets on a grim day in Aberdeen, it's down to other end of Union Street, where a local soup kitchen is about to begin serving hot food and drinks for the city's forgotten population. 'I got assaulted again,' shouts Brendan, a man in his late twenties and one of roughly 50 vulnerable Aberdonians queuing at the night kitchen at the peak of the Christmas shopping season.

'Again?' I reply. 'For fuck's sake!' I met Brendan a few weeks ago at a local church where he was receiving a free haircut. His demeanour was quite different then. He seemed calm and relaxed. Well turned out. It was there I heard about the initial kicking he received. A violent and unprovoked attack by a group of young men he says were drunk. The attack had left him with a broken jaw and facial injuries, which he appeared surprisingly pragmatic about. I remember thinking he didn't seem as pained physically as he should have, given the extent of his injuries.

Tonight, however, as the noise grows behind him, he appears fearful and brow-beaten. He is also in physical agony. He tries to lift his t-shirt to show me what he assumes will be a mark on his upper back. It's immediately apparent in how much he strains simply to remove the garment that, despite the absence of bruising, he is very seriously hurt. I assist him in removing the t-shirt. 'Arrrgh,' he yelps, with an involuntary and agonised cry. 'Did you hear that crack?'

'Yes,' I reply, 'you're fucked.' And he is. Barely able to talk due to the broken jaw he got two weeks prior and now excruciated by even the slightest upper-body movement, Brendan should be in hospital on a high dose of painkilling medication. But he will not go. He has not reported the assaults to the police either. And the reason for that is very simple: interactions with health services and law enforcement are not only difficult to initiate and sustain but could make him more vulnerable.

If he goes to the police, he's likely to be labelled a 'grass' or possibly even arrested himself. And his own dealings with law enforcement throughout the years, and those of people he knows, have shaped his view that cops aren't there to help him. That they cannot be trusted. If he goes to accident and emergency and ends up being admitted to hospital for a period, not only will his dog be left alone but his residential status in the accommodation he has recently acquired may become contested by his social landlord. With social housing in such short supply and so many homeless people looking for a place to stay, vulnerable tenants must be seen to be living in their homes or risk abrupt eviction. How indiscriminately these rules are enforced may shock you.

Earlier in the day, I spent some time in Tillydrone, north Aberdeen, with a man called Michael. Michael moved from England to work as a scaffolder on the oil and gas rigs but, like many, had fallen on hard times since the collapse of 2008. I was accompanying charity workers from a project called Somebody Cares, who were delivering some furniture to Michael from their premises – a sprawling 49,000 sq. ft warehouse. The charity, founded in 2002, provides struggling households with essential items that turn a house into a home. Individuals and families are referred by the local authority and invited to choose from a selection of items, which the charity aims to deliver within one week – free of charge.

They were on a war-footing with a fleet of five removal vans constantly being emptied and reloaded with donations of furniture, clothing, toiletries, accessories, books and bed linen. It wasn't even lunchtime when I arrived and they were on their fourth delivery, to Michael, who had just moved into his new housing estate flat after two-years of couchsurfing and rough sleeping. After a brief tour of the warehouse, which is on a scale not unlike a large department store, I mucked in with the heavy lifting and we loaded the van quickly before setting off.

On the way, Eddie, one of the senior staff at Somebody Cares and a man who has experienced his own adversities, filled me in on the recent explosion in demand for support of this nature. 'It's just nonstop,' said Eddie. 'It's like Heathrow, one lands the other one takes off, one van's in one van's out, one takes in the other one puts back out.' He tells me that in 2018, Somebody Cares supported around 4,500 households, despite receiving no financial support from the council – their biggest customer. 'So, you think of that over 16 years – that's a lot of homes we provide for.'

We pulled up outside the property, navigating some rather treacherous black ice – conspicuously absent on the pavements lining the plusher Union Street – as we slowly moved the couch, chair and a few chests of drawers into the flat. An elderly man was stood at the door, which confused me at first. 'I'm Michael,' said the man, sporting a long, dirty beard. Once the goods were in, Michael shared the story of the eviction that had left him begging on the streets. 'I went down south to visit my family, who I hadn't seen for 30 years,' he explained. 'I planned to stay for three days but ended up being there for three weeks. When I got home, I had been evicted. They said it was because I abandoned my flat, but I didn't.'

Puzzled, assuming the landlord responsible had broken the law and could be held to account, I asked Michael who evicted him, hoping he would give me the name of a private landlord or letting agent, where unethical and illegal practices are rife.

'The council,' Michael said. I was stunned.

'How old are you?' I asked.

'I'm 75.'

My jaw hit the floor. A man of pensioner age, turfed out on the street for having the cheek to stay with family for three weeks, shocked even me – someone who thought he knew rather a lot about poverty. John was unable to contest the decision that led to him becoming homeless. He was frozen out by an opaque administrative maze populated by faceless desk-killers. An organisational jigsaw puzzle where decisions with life-and-death implications are made behind a curtain of unaccountable officialdom. Yet for all the paperwork involved, the reasoning behind them is surprisingly hard to ascertain. One day, you just get told you are being evicted. The next you are sleeping on the streets – with a right to appeal. The local authority in Aberdeen very wisely operates a zero-tolerance policy where media scrutiny is concerned, making it difficult for journalists to conduct even simple enquiries unless they are related to matters of broader public awareness. It is no surprise, then, why vulnerable people dread their dealings with many public services and, like Brendan, risk their health and personal safety to avoid them altogether.

Back at the night kitchen, the queue for food is now around the corner. There's a strange party-like atmosphere. People aren't just here for the food but to socialise. There is surprising laughter and cheer. Stories are shared, barbs are traded. Injuries and wounds are compared.

'We are doing around seventy people, four nights a week,' explains Michelle, a local businesswoman who uses the basement of the fancy dress store she runs as a drop-off point for food donations.

'Seventy people every night?' I ask, thinking that number a little high.

'A thousand a month,' she says, gesturing towards the food. 'So, what you're seeing here goes really fast.' It's barely two degrees out but thankfully the rain has stayed off.

'If it wasn't for them . . . for toasties and all . . . then . . . you'd be . . .

you'd be fucked,' says Keith, another regular at the night kitchen. If not for the generosity and compassion of volunteers like Michelle then vulnerable people like Brendan and Keith would not only go hungry but they'd also be suffering even greater degrees of isolation due to a lack of social contact. This relational poverty, where those experiencing adversity become further deprived of human fellowship, creates acute isolation which is hard for even a resilient person to bear. That's why the atmosphere tonight, at the soup kitchen, is so electric – in their world, being treated with a little respect, and listened to, is a cause for celebration.

Of all the social problems administrations across the home nations face, homelessness is the least complicated. The solution is a rights-based approach to housing, where every citizen is guaranteed by law a safe, secure and affordable dwelling, and where local authorities are funded adequately to ensure the housing supply (and quality) is maintained. People will argue until the cows come home that it's a bit more complicated than that – it isn't. Not this. And don't let anyone tell you any different. Give everyone who is homeless a home. Simple. It is from this basic foundation that assessments can then be made of which problems cannot be solved by giving someone a house. The crux here is that every problem a person may be experiencing is exacerbated by residential instability, as having nowhere to live creates countless additional problems. Yet, for some reason, in the UK, despite both the clear demand for housing and the ample capacity to supply it, vast resources are instead deployed in addressing 'stigma', withdrawing money from old services to shore up new, hastily drawn-up initiatives, and, of course, endless circular crisis management.

In 2018, news that 726 homeless people died on the streets of England and Wales – the highest year-to-year increase since records began – was, much like everything else, all but drowned out by the ongoing psycho-drama of Brexit. The rate at which this vulnerable group continues to perish would have defied belief not so long ago. These days, we seem strangely adjusted to the death and despair, despite how visible the crisis has become. What the exhibition of rough sleeping taking place across Britain often obscures is the complex personal circumstances behind every visibly homeless person.

In the autumn of 2018, I encountered a homeless man named John during a Facebook Live broadcast. Of all the topics that formed the late-night, scatter-gun discussions, few drew as many viewers, or drove as much interaction, as that of mental health. John was some-one I knew of but hadn't really spoken to before. In the two-hour chat, as scrolls of people left comments and questions, John shared that he was entangled in a hellish personal crisis. Homeless and hun-gry, he had been sleeping in a tent which he had been forced to move every few days, for his own safety, until one day he returned to find it had been confiscated. On that day, demoralised, exhausted and starv-ing, John contemplated ending his life, no doubt entranced by the strangely serene allure of the River Clyde, the banks of which had been his pitch. The only reason he hadn't thrown himself in, he revealed, was that he could not bear the agonising thought of leaving behind his beloved dog and faithful companion, Oscar.

Socially isolated, with no source of income, sinking ever deeper into his depression, John was (whether he realised it at the time or not) crying out for help. Such a cry is easier to detect when you yourself have made one in the past. A few others and I acted quickly, aware that John's life was at risk. I sent him a message offering him some money to tide him over for a few days, which he refused. 'Mate, I don't have a bank account, plus I don't want to take money from any-one,' he insisted. I asked him what items he needed. He was far from forthcoming. It struck me as odd that he could be so open about his difficulties yet refuse to accept help in the very same breath. So deeply embarrassed by his terrible run of luck, he hadn't even told his friends or family he was sleeping rough. It took a while to persuade him to accept my offer.

I was in London at the time, so I coordinated with a couple of other people who had taken part in the broadcast, transferring funds to an account to be withdrawn and given to John the next day, but main-taining contact with him was extremely difficult.

We finally tracked him down to Byres Road in the heart of Glas-gow's plush West End, where John spent many of his afternoons beg-ging. I notified the other guys and they went down to meet him. John grudgingly accepted the money, clothing and food. This simple act of kindness and solidarity immediately invigorated him, temporarily relieving the toxic and corrosive feelings of shame, loneliness and acute sense of dread that can overwhelm those who have become socially excluded to the extent that he had.

John was disgusted by himself. Affronted that he had fallen so far. He was keenly attuned to how he was being perceived by members of the public, whether they pitied or despised him. He was trapped in the awful, life-threatening purgatory between desperation and pride, unable to survive without charity but deeply pained by the indignity of accepting it. He had fallen as far as a person can, in a society such as ours, yet I was struck by the fact that he never blamed anyone else for his predicament – not his family, not his friends, nor even the society which had, in many ways, washed its hands of him.

On a methadone script, having managed to come off heroin, he was struggling with the converging adversities associated with social exclusion. Estranged from his family, ashamed to reach out to friends, cut off from the benefits system due to having no address and surrounded by offers of alcohol and drugs, rough sleeping was the easiest and simplest part of the day for him. This level of exclusion is like being encircled by a deep moat with no way across; interplay between the structural barriers of having no address and no bank account, and the various problems of proximity, whether being too distant from family or too close to hard drugs, creates excruciating living conditions which are only tolerable for so long.

Beyond the personal circumstances of every person who becomes homeless and is forced to sleep on the streets there are many other factors that can make life even harder. Not least the weather conditions. The same weekend that I made contact with John, three bodies were fished out of the River Clyde within a 24-hour period. The spate of apparent suicides took place during what the Met Office described as a 'cold snap' – a sudden drop in temperature, often compounded by a bitter wind and sleet or hail. While there is no way to confirm why three people decided to leap to their deaths in the freezing river that weekend, I had my suspicions that such unforgiving conditions may have played some role.

It can be hard for the average person, with no frame of reference for this kind of social exclusion, to conceptualise what that sort of desperation must feel like. Being literally locked outdoors, with an empty stomach, as hail, snow, sleet and high arctic winds blow in around you. In times like that, alcohol and drugs are often what keep people alive, thanks to their physically and emotionally anaesthetising effects, though I suspect being unable to obtain them, during a cold snap, may act as an additional form of propulsion into suicide.

I became determined that John wasn't going to be the next soul to expire in that cold, black river. Confident his situation was stabilising, I made a public appeal for support. Most people who responded did so to express their admiration of his strength. Respect for his resilience and awe at his courage. Many also voiced their anger that so many of the services set up to support people like John had seemingly failed him. However, a few, upon hearing John's story, saw his misfortune as an opportunity to either broadcast their ignorance around the issues he was facing – many pointing out that he couldn't be that hard up if he was able to look after a dog – or to make glib, self-serving points about why immigrants were to blame. Ironic, perhaps, given that John's own social media activity consisted mainly of posts detailing how much he detested far-right figures like Tommy Robinson and Katie Hopkins. Figures he believed were using the struggles of people like him to cultivate unjustified anger at asylum seekers, refugees and immigrants.

Increasing numbers regard the homeless and destitute as a nuisance. They view beggars and the homeless as social parasites undeserving of compassion, whose adverse circumstances are the result of poor choices. While many delude themselves that such a reaction is instinctive, it is not a natural response to human suffering. In truth, many adopt such uncharitable attitudes following an initial feeling of empathy or concern, which is followed quickly by irritation and anger when confronted by a homeless person whom they feel utterly powerless to help. This powerlessness can register as agitation and fester as resentment. Adopting a hostile attitude towards a vulnerable person is often a way of closing off the much more natural sense that we may have some obligation to help them. It's not one individual homeless person many of us will come to resent but increasing numbers, year on year, in the alleys, stairways and pavements we traverse going about our own, stressful lives. As the problem presents on the high street more starkly than ever before, our prior kindnesses seem to have amounted to nothing. Offering still more compassion, or generosity, begins to feel a thankless and futile waste.

We all have problems. Our empathy is finite, dependent on our mood and what is going on in our life at any given moment. When we feel powerless in the face of the immense injustice on display, it gradually becomes natural to switch off. Some of us are able to observe this impulse to shun the desperate and self-correct. Others may attempt to justify their momentary anger as a natural aversion to the weak, the vulgar and the undeserving. What we all have in common is that we are disgusted by what we are seeing daily on our high streets.

From the point of view of the rough sleeper, our avoidance often registers as indifference, creating a painful sense of invisibility. Hundreds, maybe thousands, will walk past someone begging before somebody decides to throw some change their way. For the person sleeping on the street, every potential interaction can become a game of Russian roulette. As among us, there are some disturbed and cruel individuals for whom abusing the homeless population verbally and physically is, apparently, a great laugh.

What is most worrying is how this public hostility extends to the environment itself, finding expression in an increasingly callous and twisted urban landscape. Across Britain, the real innovation, where homelessness is concerned, is not found in our approaches to supporting rough sleepers. Our real ingenuity is expressed in the many inventive, state-of-the-art methods we employ to deter them from public spaces. In different towns and cities, you'll find countless examples of 'hostile architecture' – an intentional design strategy meant to regiment or restrict behaviour. Also known as hostile design, unpleasant design, exclusionary design or defensive urban design, hostile architecture's most enduring achievement thus far has been the creation of 'anti-homeless spikes' – metal studs embedded in flat surfaces to make loitering, begging or rough sleeping dangerous or uncomfortable. Other methods include sloped windowsills to stop people sitting down, benches with the armrests positioned specifically to prevent people lying on them and even water sprinklers that activate intermittently, without warning.

Hostile design directly targets those who rely on public space more regularly than others and represents something of a boon for up-and-coming product designers and architects, as rough sleeping in the UK continues to rise and solutions to the problem of beggars and the homeless are sought. This deliberate and callous form of social engineering works in harmony with public apathy and structural class barriers, creating the desired effect: we blame the homeless for being a nuisance, loitering in places they need not be, despite the fact we make it our business that they have literally nowhere else to go.

By November, John was doing much better. The response to the public appeal was overwhelmingly positive. He was offered a place to stay for a few nights, where he was able to have a shower and sleep in a proper bed. Someone else took him in and let him stay at their flat, as they were usually out, caring for an elderly parent. For weeks, I received countless offers of support, which I immediately relayed to John. He took further strength and encouragement from the sense that he was no longer alone. That he had no reason to be ashamed. A little solidarity can go a long way to breaking the power of depression. Feeling supported, or that you have a voice, can often bring about the game-changing epiphany that breaks a soul-crushingly negative and potentially life-threatening mood.

It was not long before John marshalled the strength to begin pulling himself out of the swamp of poverty, towards the dry land of recovery. Now with an address, he applied for benefits and began looking for work. Towards the end of the year, I invited him to a gig at the Barrowlands in Glasgow, a prestigious music venue where I and some other artists were raising money for and awareness of Glasgow's homeless population. John was understandably anxious, having spent so long in social isolation, prolonged periods of which can leave an individual extremely daunted by human contact.

Isolation plays a key role in partitioning a vulnerable person from the support they need, irrespective of its availability. The longer someone goes without social connection, the likelier they will begin to experience mental health problems. Indeed, addiction itself arises from a desire to connect. As painful as loneliness can be, for many who become isolated, it offers an alluring, albeit illusory, predictability – especially those struggling with addiction or mental health problems. The strange pull of a dark, empty room. The quiet call of a deserted stairway. The familiar thrill of an empty bus. When you have experienced a problem, helping another person through it becomes a matter of intuition, so I tried to make it as simple as possible for John to attend the gig by offering to cover his taxi fares there and home.

At the venue, just before I was about to go on stage, I finally got to meet him. It was an emotional experience. We embraced and I commended him on how well he was looking. He wanted me to know how grateful he was for all the support. Support which, to me and others, required very little effort but to him had clearly meant a great deal. He seemed fresh-faced and optimistic. He had finally embarked on the terrifying and challenging journey back from the margins and into 'society'. His sense of hope had been restored by the great many people who had extended love and compassion to him, unconditionally and without judgement.

We made plans to reconnect after Christmas and I left him to his first night out in years. When I got home, I noticed a message he had left me in advance of the gig: 'Mate, it would be cool to catch a chat with you for ten minutes to thank you for giving me what I needed to now have my own temp furnished flat,' he wrote. I was elated that a small act of love could still go a long way, in a world where it too often seems in such short supply.

John was one of the lucky ones. He had made it. But not even two months later, I was staring through bitter tears, in utter disbelief, at his coffin. John now lives on as a statistic. A statistic that could never truly convey the extent of his struggles and experiences, nor the dumbfounding efforts he went to in his struggle for basic human dignity. A statistic that renders insignificant everything he endured before his life was cut so needlessly short. After seemingly making it out of the quicksand of absolute poverty, he was snared by 'street-Valium' – one of many counterfeit substances which are driving the drug crisis in the UK.

I sat there in the church in the East End of Glasgow, across the aisle from his parents, who were inconsolable as Judy Garland's 'Somewhere Over the Rainbow' played their boy out for the final time. I too wept as the curtains drew slowly to their agonising and inevitable close. I weep now as I write these words. Attending funerals with increasing regularity is, for many of us, nature's way of reiterating that the curtains are always closing in around us. That time is life's most precious commodity and mustn't be allowed to dwindle, like sand between our fingers, through soft and careless living. Sadly, for too many on the sharp end of social inequality, funerals are rarely about celebrating long lives well lived, but about mourning yet more senseless and preventable death.

Drug problems are sadly synonymous with this kind of extreme, residential instability. Figures published by the Glasgow Health and Social Care partnership in 2018 revealed 17 occasions over a four-month period when a person who had used drugs was revived at Glasgow's Winter Night Shelter. I could rattle off a raft of grim and shocking statistics but in the context of a drug-death crisis in Scotland the facts do little to move the needle in any direction; drug-related despair and death (which we will examine later) have become so commonplace, we now rarely bat an eyelid unless it's someone we know.

Having spent so long without a home, in chronic isolation, trying to solve the jigsaw of public services, contending with apathy and abuse from the public, while traversing an urban landscape configured to repel him, the odds were always stacked firmly against John. But despite the clear adversities he did experience, his situation was actually far more straightforward than that of most other homeless people with drug problems. For one, he didn't have any children or a partner. He was already on methadone so wasn't trapped in the withdrawal cycle that forces so many heroin users into crime and health-risking behaviour. He didn't come from an abusive background. I suspect that had he been offered support earlier in his downward spiral, John may have been one of the simpler cases to sort out. I'm certain that he had no intention of dying. That his death was a tragic accident. One that could have been prevented had services been equipped to support him sooner.

A number of forces outside John's control eventually contributed to his death. There was the pivotal role played by the various authorities and agencies which were apparently responsible for his safety and care. The police officers who confiscated his tent, plunging him into thoughts of suicide. The welfare system that is no longer capable of interpreting complex, individual circumstances due to reforms which are designed, ultimately, to discourage people from asking for help. Yet the dysfunction present in John's life, and indeed in the lives of many of his misfortunate peers, which is so often the focus of discussion around issues like homelessness, is eclipsed entirely by the dysfunction exhibited by local authorities – in this case, Glasgow City Council, Scotland's most decrepit institution.

Glasgow, like every UK city and town, behaves duplicitously where its socially excluded citizens are concerned, the most acute form of social exclusion being that experienced by rough sleepers with drug-problems. Across the UK, the number of rough sleepers has surged in recent years – falling slightly during the pandemic when government acted to prevent the spread of infection by placing many people experiencing homelessness in temporary accommodation – due to a lethal combination of budget cuts, soaring rents and a chronic shortage of affordable homes.

In the year leading up to John's death, the various public bodies, agencies and charities which exist to help the socially excluded were slightly distracted. In fact, they were openly preparing for a war – battening down the hatches and rationing their resources and services. This war of attrition was not waged on them by an economic downturn or some unforeseen climate catastrophe, but by the UK government, in the guise of welfare reform. Reforms which, much like hostile architecture, were symbolic of a shift in Britain's posture towards the vulnerable and destitute. In this war, local authorities are the foot soldiers, charged with the duty of passing on the exacting cost of austerity to British citizens.

While councils across the country vary in their determination and desire to tackle rough sleeping, every local authority in Britain, irrespective of any political will within specific councils to resist or face-down the problem, has effectively had its hands tied by government commitments to draw down public spending. Why the axe falls hardest on the poorest is not difficult to understand. When local authorities are deciding which services to downsize or cut, they usually find it easier – and less politically costly – to swing the axe at services that socially (and politically) excluded people depend on. The logic is simple: people who vote and pay tax are the priority for most politicians, as a matter of electoral necessity. The socially excluded have little to no representation, are usually inactive politically and economically and are, therefore, easier to throw under the bus.

While austerity has rightly become synonymous with the Conservative Party, if you pay closer attention, you'll find that visiting unjustified hostility on the poor and vulnerable is not the exclusive preserve of Tories. Indeed, in Scotland and across the UK, local and national administrations have sought to contrast themselves with the Conservative government rhetorically while playing the same political games as they do – by pandering to public ignorance around begging and rough sleeping. In 2016, Labour-led Nottingham City Council was forced to withdraw an anti-begging advertising campaign by the Advertising Standards Authority. The posters, dotted around the city centre, prominently featured slogans like 'People who beg often have serious drug or alcohol problems. Please give to charity, not to people who beg.' Others read 'Begging: watch your money go to a fraud' and 'Watch your money go up in smoke'. The authority said the language used in the adverts 'portrayed all beggars as disingenuous and undeserving individuals that would use direct donations for irresponsible means'. It added: 'We further considered the ads reinforced negative stereotypes of a group of individuals, most of whom were likely to be considered as vulnerable, who faced a multitude of issues and required specialist support.'

This view was echoed in Glasgow when, in May 2016, Robert Aldridge, chief executive of the charity Homeless Action Scotland, condemned the launch of a begging survey conducted by Community Safety Glasgow, one of Glasgow City Council's many arm's-length organisations. Disturbingly, the questionnaire asked residents if they had been affected by 'aggressive begging' and, if so, how would they like it dealt with. Similar surveys have been conducted across the country, representing a distinct shift in both emphasis and accountability where our most vulnerable citizens are concerned. It has become much easier for politicians to placate public resentment and anger by feeding misconceptions and prejudices around beggars and the homeless than it is to actually help them. These surveys are simply an attempt by two-faced local authorities to ascertain precisely how much they can get away with not doing about the issue. For if enough of the public are seen to take an unsympathetic view of rough sleepers and beggars, it lets the council off the hook, politically.

Speaking to the *Herald* newspaper at the time, Aldridge said: 'This survey asks many things about people affected by begging and beggars but in our view is asking leading questions which will skew the survey responses in favour of those that are negative about begging within Glasgow.' He added: 'There are a myriad of reasons why a person will beg, however what is true for all who have to beg is that they are living a life of poverty and destitution, with bleak prospects for the future. They are in need of support not vilification.'

Local authorities have a statutory obligation to place people experiencing homelessness in suitable accommodation. These institutions break the law hundreds of times every day by failing to do so – yet are subject to no penalty. What incentive do they have to get serious about rough sleeping and begging when inaction which results in misery and death goes unpunished? Those working in the sector, across local government and within charities, are well aware there's plenty of blame to go around but telling home truths about this stuff can make you unemployable. This concern often takes precedence in the minds of individuals who may otherwise speak frankly. The cumulative effect of individual professional self-interest and the corresponding aversion to rocking the boat is an institutional culture of resignation and tacit complicity, which reverberates down the social scale gradually, expressed statistically as death.

In criticising the various institutions which preside over the mayhem, you risk implying that those who work within them are uncaring. That they are not trying. This is absolutely not the case. Indeed, it is the fact that so many across this sector desire better outcomes but are unable to produce them which speaks to the central problem – institutions themselves. Homelessness is always framed as a wickedly complicated problem with no easy solutions. That may be partly true. But be in no doubt, the repeated institutional failure at every level of governance across the UK, and the lack of accountability in the face of it, lies at the root of this crisis. A crisis that shames Britain.

6

Addiction

'Years ago, when I started selling it, I wouldn't sell it to a woman if she had a kid in a pram,' a middle-aged drug dealer tells me, within spitting distance of his MP's constituency office in a Glasgow housing scheme. Like many dealers, this gentleman made his initial incursion into the grim, morally ambivalent world of drugs as a low-level supplier after losing his job. His employer, Scotrail, sacked him when he refused a drug test, which led to the repossession of his home and the eventual breakdown of his marriage. 'Within a year, I was standing here selling heroin,' he says.

He started like many dealers do, turning a modest profit by purchasing relatively small amounts from a bigger player to supply the local demand in his area. However, as is often the case, he succumbed to the curious allure of the substance he was peddling.

'Are you not aware of the damage the drug causes?' I ask, perhaps glibly, keen to hear how he now justifies selling it to vulnerable people, like young mothers and the homeless.

He pauses briefly, locating what, to him at least, feels like an appropriate, face-saving response: 'I never took heroin until I was 27. I went through a heavy withdrawal.' The withdrawal associated with heroin and, indeed, all opiates, is brutal and unforgiving. Serious heroin addicts with a high tolerance for the substance become trapped in the hellish cycle of scoring drugs to escape withdrawal symptoms, which isn't cheap. This is why criminality becomes a necessity for many. This bone-chilling fear of withdrawal trumps the shame of social exclusion, the indignity of prison and even the permanence of death. The need to pursue the next fix, at the cost of all else, leaves many drug abusers physically and morally disfigured. Only when contending with that reality does my 50-year-old friend's rationalisation for peddling smack sound a little more plausible. Having gone through a 'rattle' – slang for withdrawal – he, in his own words, 'knew how much these people needed it'.

It's broad daylight in the heart of the bustling post-industrial community of Possil, north-west Glasgow. Another two men loiter behind a pair of battered phone boxes which partly shield them from the glare of a busy high street. They're here to score. It's not even tea-time and nobody anywhere is batting an eyelid. The police station round the corner will close shortly, not that it will make a difference. Here, even the cops have all but accepted that drugs and their attendant miseries are inevitable.

Scotland has a proud pioneering history. We brought the world antibiotics, telecommunications and a paradigm-shifting Enlightenment – which we never shut up about. But we also lead the way in delusional hubris, perceived victimhood and reflexive nationalist finger-pointing – cultural defects which we are less willing to lay claim to. In Scotland, our successes are our own but our failures – and they are many – are usually someone else's fault. And by far the greatest blemish on our storied history in recent times is our abandonment of the most vulnerable, persecuted and excluded group in our society: people with drug problems.

Data published by the Scottish government in 2019 revealed that in Scotland, drug-related deaths rose by a shameful 27 per cent to 1,187 during 2018. That means the death toll in Scotland was equivalent to five Lockerbie bombings or fifty 7/7s. It was nearly three times that of the UK as a whole and, per capita, the drug-death rate in Scotland was higher than that of the US. Yet no national emergency was declared. The drug-death figure increased again in 2019, by 6 per cent, according to official statistics released the following year. National Records of Scotland figures revealed there were 1,264 deaths, an increase of 77 on the previous year and the highest figure on record. Yet still, no national emergency was declared. By contrast, in 1996, an *E. coli* outbreak, which affected 200 people and caused 21 deaths, led to a high-profile public inquiry which produced a damning report placing blame on a Wishaw butcher and council health officials. But as the drug-death numbers rocket skyward, with the equivalent of a large football stadium full of people managing substance misuse issues, there is, much like the homelessness crisis, a deathly dearth of accountability.

The latest statistics, published by the National Records of Scotland in 2021, revealed 1,339 people died in drug-related deaths during 2020. Scotland has by quite some distance the highest drug-death rate recorded by any country in Europe. While the causes remain disputed and continue to baffle many, people from the poorest communities are 18 times likelier to die a drug-related death than those from more affluent areas – a damning statistic.

Before the Second World War, Possilpark was a bustling working-class community and a stronghold of Glasgow's industrial juggernaut. The area developed economically around Saracen Foundry – an iron-works which served as the area's main employer. Possil's dependence on this growing industry is evidenced in the steep population growth that occurred in the late nineteenth century when it exploded from just 10 people to 10,000 within a 20-year period. Many of the city's decorative centrepieces, including railings, water fountains and bandstands, were crafted by Saracen Foundry, which shipped its products around the globe. After the war, traditional industries fell into decline and Saracen Foundry's closure in 1967 marked the dawn of a new age of mass unemployment, poverty and crime, as a heroin epidemic loomed.

This is a familiar story across the UK. The mechanisms by which working people once lifted themselves out of poverty were wound down or closed abruptly, and with no strategy in place to ensure a just transition, communities built around these traditional industries became economic wastelands.

The drug problem is worse in Scotland for the same reason the drink, cigarette and life-expectancy problems are worse: the severity of deindustrialisation coupled with a lack of proximity on the part of the UK government prior to devolution set this ghost train in motion. These issues are, in my view, that much more acute north of the border, in part due to how radically Scottish society was transformed by the industrial age and the concurrent lack of political autonomy required to develop an alternative economic strategy necessary to meet the challenges of the post-industrial period. This dynamic is mirrored in aspects of drug policy itself, some of which remains devolved to Westminster.

Exacerbating the economic factors historic to areas where drug deaths are highest is the lack of investment in support services from the Scottish government since 2013. In some cases, the Scottish government even cut funding, despite warnings it would likely lead to an increase in deaths. For this myopic neglect, thousands of families have paid the heaviest price.

An additional compounding cultural factor is the panic that grips communities of drug users who can't go a day or two without hearing another story about someone who's died, yet, due to severity of their own addictions, and low accessibility of services, cannot stop using. Thanks to a largely ambivalent mainstream media (with a few notable exceptions, not least the *Daily Record*, which has made drug reform a central campaigning issue), people with drug problems have invariably been portrayed as vulgar, selfish and feckless individuals whose problems are entirely self-generated. The notion that they may actually care about each other and, indeed, take care of one another, has, thus far, eluded many. But drug takers survive by moving in tight-knit communities, bonded by the acute and constant social exclusion they endure – much like those experiencing homelessness. Every time one addict is found dead, others grieve. They sense too that they might be next. They feel that society has all but abandoned them. Socially excluded, bastardised by the public and misunderstood by many who wish to help them, what is there to live for? And so, they do what we all do when terrible things happen: they reach for whatever numbs the pain. As they do, the social contagion spreads, tearing through entire families and communities, leaving death and despair in its wake.

Back on the street corner, as the men candidly regale me with their tales of drug addiction, the addict in me begins to feel nostalgic about my past drug use – a sign of the insanity central to the condition of addiction. Just the sight of a drug, in the absence of a strategy to deal with it, can lead to the complete highjack of an addict's executive function. Before they know it, they may be using again, and won't understand fully why. Those who have never experienced what it's like to use drugs against their will are naturally incapable of conceptualising such a predicament.

The central problem for the addict is the recurring idea to acquire and use drugs no matter the cost or consequences. An addictive obsession induces in the mind of a drug user a carousel of revolving delusions about the nature and true extent of their problem. And these delusions go beyond the realm of mere false belief and are in fact the result of neurochemical processes much like appetite or sleep regulation. Once the thought is in, and without the tools to deal with it, it becomes very hard to resist.

While addiction presents outwardly as a result of free will (addicts 'choose' to keep using drugs despite the negative consequences) there is an involuntary component to how the addict thinks and behaves. Addicts tend not to perceive the choices which are truly available to them and addictive compulsion impairs how they then assess and select from the narrow range of false choices they do perceive.

The addict will move through them as part of a process, which usually begins with the idea they can safely control and enjoy their substance – or some variant of that basic idea. Having imbibed it, they then lose the ability to regulate either how much they subsequently take or how they behave under the influence, which then requires further denial and self-justification to sustain. Eventually, the addict begins to recognise the gravity of the situation but also becomes resigned to the false belief they cannot stop. It's only when the addict reaches the final stage, when the magnitude of what they are suffering from, and the cost to them and others is too great to be denied, that they become once again susceptible to the truth and that the window of opportunity for intervention opens. This often requires a shocking event: a loss of job or liberty, or something which temporarily interrupts the supply of alcohol or drugs. It is then and only then that an effective intervention can be staged. One which is not simply about the catharsis of concerned friends or loved ones, or the aims and objectives of a government or service, but about the mental, physical and relational needs of the addict themselves.

Intervention is about timing and circumstances. Is the person ready? How much do they feel they have to lose by not getting sober? Who do they have around them and what resources do they possess emotionally and materially to truly help? One of the biggest problems we have culturally is this cross-institutional misunderstanding of addiction as ultimately being a choice, which has led to a drug support infrastructure which is often misaligned with the needs of problem users.

This culture gap between drug users who have slipped into addiction and people who have never tried drugs, or who use them safely, creates an additional gulf in which misguided preconceptions about the phenomenon of addiction and, indeed, the plight of addicts emerges. What people often don't understand is why so many are attracted to drugs in the first place. The answer is simple: drugs make you feel really good.

I drift into a daydream about that half temazepam from years ago that I should have just swallowed before the police got their hands on me. 'What a waste,' I think to myself, able to recall its sweet taste and chalky texture in my mouth. Also known as eggs, norries, rugby balls, moggies, mazzies or downers, tranquillisers come in different forms. One is a small egg-shaped pill, soft to touch, not dissimilar to the texture of a Kiwi fruit. The other, a thick, white, chalky tablet. They start to take effect between about 20 and 40 minutes after swallowing, at which point a euphoric lethargy washes through mind and body. You are not tired but should you close your eyes, you may appear to be asleep. Falling asleep after the drug has taken hold, though, is a terrible waste of its pleasant effects. Staying awake is also important because it can be terribly hard to wake up out of a tranquillised sleep. Not a smoking chip-pan nor burning curtains could stir you from a sleep like this. And physically getting up is quite the chore, whatever reason forces you to do it. Limbs are heavy against whichever surface they are rested. A hard floor becomes a soft bed. The constant physical tensions you often mistake as your natural expressions (a furrowing brow, a tapping toe or an incessant nail-picking finger) subside, joining an exodus of unintentional scowls, facial twitches and navel butterflies. You feel heavy but the extra weight is offset by the lack of desire to move. Should you wish to speak, you may notice your voice noticeably quieter – not because it is being suppressed but because it returns to its natural level, no longer intensified by the stress placed on the voice-box. Words come a little easier. You listen more intently. You become more considered and sensitive in your responses. You may also feel suddenly capable of articulating precisely that which you so often struggle to express. Words which previously eluded become your keen and humble affiliates. These drugs are attractive to many for a reason – they induce a peace, serenity and a profound, value-shaping sense of wellbeing, which is difficult for even the most articulate drug users to describe.

But the drugs I describe, which flooded Scotland's poorest communities 30 years ago, were not the vulgar concoctions of illicit dealers but pharmaceutical-grade, state-approved chemical compounds distributed by physicians and pharmacies. These drugs were prescribed so widely, in the wake of the heroin epidemic, that some addicts have been chasing their effects ever since.

Drug abuse is so often framed as an issue for the feckless individual but we must also consider how thoughtlessly these drugs are often introduced into society. Did it not occur to anyone that flooding communities like Possil – known primarily for its stress-related health problems – with powerful psychoactive substances which relieve stress might not be the best idea? What risk-assessments were undertaken before benzodiazepines and tranquillisers were introduced through the NHS? When drugs are road-tested for approval, their broader social implications are rarely contemplated, despite historic evidence that their marketing campaigns are often a little misleading and that medical professionals, while highly qualified, often lack the proximity to poorer communities which is so vital to understanding how an increased availability of mood-and-mind-altering drugs might play out, socially.

In 1898, the German drug manufacturer Bayer marketed diacetylmorphine (an opiate) as an over-the-counter cough suppressant under the brand-name Heroin. Heroin was developed as a morphine substitute but, despite being marketed as 'non-addictive', it wasn't long before it scored some of the highest rates of addiction among those who used it. Fast-forward to the opioid crisis in the United States, where over-prescription claimed the lives of more than 200,000 Americans between 1999 and 2016. The United States is home to 5 per cent of the world's population but accounts for around 80 per cent of all opioid use. Again, those communities devastated by deindustrialisation are the ones more adversely affected, with the opioid crisis rising sharply in the period of economic decline between 2001 and 2009.

Where the US led, Scotland followed. In 2018, the *Scotsman* reported on research published in the *British Journal of Anaesthesia* by researchers from several Scottish universities which found that 18 per cent of the Scottish population was prescribed at least one opioid pain-killer in 2012. They found that 'there were four times more prescriptions for strong opioids dispensed to people in the most deprived areas, than to those in the most affluent areas'. All data available points to a large increase in the prescription of addictive painkillers like co-codamol and tramadol over the last ten years or so, while other substances have been restricted, driving demand for counterfeit equivalents.

Much of the frustration felt by those who have lost loved ones or who have battled addictions themselves arises from a perceived lack of humility and honesty about the role the drug sector and health services have played in bringing us to where we are. In 2009, according to the National Records of Scotland, benzodiazepines were implicated in, or potentially contributed to, 154 (28 per cent) of the deaths but by 2020, there were 879 deaths linked to so-called 'street benzos' – demand for these drugs sky-rocketing when Scottish NHS providers suddenly restricted benzodiazepine prescribing, forcing many into the deadly embrace of the unregulated black market.

The drug crisis has been wrongly framed as an issue about danger-ous cocktails of drugs but many of the illegal substances on the black market today are simply amateur attempts to mimic the effects of prescribed substances which were either withdrawn or restricted by government.

Many of the drugs dispensed in the 1990s, particularly benzodiaz-epines, were prescribed as part of a strategy often referred to as 'harm reduction'. This is a pragmatic government response to widespread, serious drug use and encompasses treatments, services, policies and public awareness campaigns that aim to reduce deaths, overdoses and bloodborne infections associated with intravenous drug use, as well as social and cultural factors such as criminality and stigma.

Today, the most widely prescribed harm-reduction treatment for heroin addiction is methadone, an opiate substitute therapy (OST), the prescribing of which increased substantially in the UK in the 1990s. OSTs provide a range of benefits for the individual user and the wider community in that they suppress the effects of heroin withdrawal, thus allowing people with a serious, long-term dependence on the drug to cease heroin use safely. The effects of OSTs also last longer than a heroin fix, meaning the sufferer is not trapped in a cycle of having to acquire and use drugs every few hours. And with illicit drugs removed from the equation, the chances of overdose, poisoning or infection falls significantly, as do incidences of drug-related crime.

However, in recent years, OSTs have become the subject of fierce debate as drugs like methadone are also implicated in drug deaths, usually as a result of the user taking other substances in addition to what they have been prescribed, which, for many, begs an obvious question: if methadone is sufficient to get people off drugs why are they using other drugs, or drinking, on top of it?

A harm reductionist might tell you (and in many cases they'd be absolutely right) that the person who relapses or dies may not have been receiving an 'optimal dose' – enough methadone to supress the effects of withdrawal. But many with experience of addiction would likely argue that the condition cannot be managed long term with an 'optimal dose' because a key feature of the phenomenon of addiction for some is that no amount of drugs is ever enough. Basically, that the concept of reducing harm by replacing one drug with another drug, long term, is fatally flawed because it is rooted in the misconception that an addict can control their appetite for illicit drugs while still dependant on licit ones.

If I have a history of crashing whenever I get behind the wheel of a car, will a different model of car reduce the risk of an accident or would it be wiser that I go back and do some more driving lessons? With that analogy in mind, let me emphasise this point for those with low proximity to this issue: if people end up on methadone because they cannot regulate their heroin use, then why is it assumed an OST, which is just as addictive as heroin, will produce the sudden ability to control their appetite for oblivion?

While this view is often dismissed by many in the harm-reduction sphere, who claim such a view is not evidence-based, harm reductionists themselves draw from a narrow evidence base which is primarily focused on treatments which involve replacing a harmful substance with a less harmful one. As a result, a culture has developed which is arguably less concerned or knowledgeable about other approaches which do not involve prescription drugs, and treatments which involve reducing or abstaining from all mood- and mind-altering substances, including prescriptions, are viewed with scepticism.

The same medical establishment which has overseen the dominance of OSTs like methadone has also, itself, expressed concerns about methadone from the inception of harm-reduction strategies. Research published in 2009 in the *British Medical Journal* refers to a 'confidential inquiry into methadone related deaths carried out in Scotland in 2000' which analysed the 18 per cent year-on-year increase in the number of prescriptions for methadone since 1996. The research noted that '56 drug related deaths took place in 2000 and were referred to the inquiry' and that in 30 of these cases (54 per cent), methadone was cited on the death certificate. In the citations, the link to the confidential inquiry referenced leads to a blank page but this observation is highly notable given that just under half of drug-related deaths in Scotland 18 years later, in 2018, also involved methadone. It is not unreasonable to ask why this OST, implicated in around half of all drug deaths, remains the primary treatment for opiate addiction.

If methadone is taken as advised, under supervision, and regular contact occurs between the patient and health services, which aims to supplement OST with structured psychosocial support, then the benefits of it are undeniable. The problem arises not from the treatment itself but from the context into which it is introduced. Addicts' lives are often chaotic and, as many form tight-knit social connections with other drug users, their exposure to illicit drug use often remains high even when they themselves have chosen to abstain. While methadone suppresses the physical need for heroin, it does not deal with the void left by its absence; people take heroin because the intense euphoria it produces provides a temporary escape from their problems. This is the aspect of addiction that poorly resourced harm-reduction strategies and initiatives are often not as effective at managing. Methadone reduces the risk of overdose, death and other harms associated with heroin, but these gains are contingent on other variables beyond the control or, often, capability of services.

When people take drugs on top of OST, when contact with services is irregular or disrupted or no psychosocial support is available, methadone's utility becomes more questionable. The more widely it is prescribed the more it is found on death certificates – not because it is dangerous in and of itself, but because when the emotional needs of people with drug problems are unmet, they become likelier to supplement OST with other substances.

When those who work in the drugs sector assert the merit of one evidence-based treatment over another (for which ostensibly there is less or no evidence), that is not always an indication of the latter's inefficacy; it can instead reflect the government's decision to devote resources to the research of one form of treatment and not another. That health professionals understood more about optimal methadone dose ranges in the 1990s than they did about the vital role psychosocial support must play, demonstrates this. And yet, anyone who had overcome their addiction through an abstinence-based mutual aid group where no professionals would have been present – as many have done – could have told them that. Whether they would have been heard, however, is another matter. The fact that countless people learn from other addicts, and not doctors, how to live meaningful lives without mood- and mind-altering substances, and become productive, valued members of society, is not regarded by some in the harm-reduction field as admissible evidence.

As with the attainment gap in education, which feeds into the labour market, criminal justice and housing, the drug crisis arises at the confluence of historic, intersecting inequalities. Serious drug users usually begin their social exclusion at school, become subject to early police involvement, perhaps even doing multiple spells in prison with periods of unemployment and residential instability in between. Without serious action on wider social inequality, drug-related death and despair are inevitable.

University West of Scotland researchers Iain McPhee, Barry Sheridan and Steve O'Rawe published findings in 2020 that within the poorest areas of Scotland, drug-related deaths were comparable with heart disease. Twenty per cent of the total number of drug-related deaths in 2017 and 2018 were people who lived in the bottom 5 per cent of most deprived areas. Given the government's commitment to evidenced-based harm reduction, you'd have thought it might have hit them by now that the best way to reduce the harms associated with problem drug use is the stop dancing around the fundamental problem – structural class barriers.

The systemic nature of Scotland's failure in this area was brought into sharp focus in 2021, when another drug death made the headlines. Three days after discharging himself from a rehab, a gentleman (whose family requested he not be named) was found dead at his home. He had checked out of rehab because he was threatened with being made homeless. Due to a fatal loophole in Scotland's ludicrous system of funding, which was rectified after the tragedy, many rehab beds were funded by housing benefit. This created a nightmare scenario where vulnerable people were forced to make the most hellish choice imaginable – you can either have your home or your place in rehab funded, but not both.

This is all the more infuriating if you understand the nightmare that awaits most people seeking residential treatment, which, like funding to other services, has been cut savagely, making it far harder to access.

For many years, a pervasive belief existed, evidenced by the savage cuts to residential treatment facilities, that rehabs (promoting abstinence-based recovery which involves the cessation of all mood- and mind-altering substances) were not as effective as harm reduction. But this only revealed the lack of proximity many in government and the drug sector had to the issue. Residential treatment, as with opiate-replacement therapies like methadone, has become a proxy in the cold war raging between those who believe abstinence-based recovery is being under-sold and harm reductionists embedded within the government's embattled drug sector.

Rehabs offer three features which no harm-reduction treatment can replicate. Residential treatment takes people with drug problems out of the dangerous, hostile and chaotic environments, which not only drive their drug use but also where drugs are readily available, and places them in a safe and supportive place, far from the toxic social connections and attachments which impair their progress. Once admitted, they can be medically detoxed, under supervision, from most drugs, including heroin and methadone. In residential rehabs, where clients live together in close quarters, the vital psychosocial support which is often lacking in individuated OST programmes dispensed through pharmacies, is on hand in the form of full- and part-time staff as well as, crucially, other addicts who are a little further along the road. This provides someone with a drug problem, who has come to believe there is no solution, a profoundly tangible example of what early recovery looks and feels like. By identifying with others' experiences, the sense of isolation which thwarts an addict's attempts to confront their problem starts to ebb away and they begin to conceptualise what a life free of drugs might look like. Once new attachments are formed, a sense of security and hope develops. As their minds slowly begin to turn from their own immediate problems towards others in treatment (as well as the various household responsibilities they will be given gradually throughout the course of their stay), the obsession to use drugs, as well as the physical and psychological cravings, gradually fades.

The additional strength of rehab is that most of the staff will have direct experience of recovering from serious, long-term addiction. Their proximity to the issue is the greatest tool in their armoury, not least because it authenticates them in the eyes of those on the lower slopes of sobriety. Indeed, their entire approach is informed not simply by their intellectual grasp of the subject at hand but by their visceral understanding of how to approach clients at their most fragile. How to break down the walls of denial with humour, sensitivity, and compassion. For them, helping people get sober is not a matter of knowledge but one of intuition. They know when to challenge you and when to leave you alone. They know when you are being sincere and when you are being economical with the truth. And they know when you are being emotionally honest and when you are performing. Staff in a rehab will see all of this in you, and more, not because they know you but because they know themselves. When the specifics are stripped away, every addict is the same – we have a head that tells us we can take drugs and a body that can't control how many drugs we take once we start. In rehab, you will receive not just a medically supervised detox and any additional pharmacological support you require, but also an optimal dose of love. You will learn to confront the true root of the addiction, which is the inability to sit with your emotions, to deal with stress or to navigate relationships – the drug use is the symptom of that deeper problem.

What this all boils down to is that, in so many cases, the essential information an addict requires to transcend addiction completely can only ever be truly imparted by another who has walked the path before them. A lot of professionals get upset when they hear such a thing said – this doesn't change the fact it's true.

A local chemist dispensing methadone daily, or an alcohol and drug support service that closes at 5pm, simply cannot provide the vital psychosocial dimension of treatment to which the evidence-based research repeatedly points to as essential – that is why methadone is found on so many death certificates year after year. The central problem with the harm-reductionist model in Scotland is not the concept of harm reduction itself. It is that in the absence of the additional support and services required to address the root causes of addiction, and the reform of drug laws which trap addicts in the criminal justice system, people either risk reverting back to their old ways or remaining on opiate replacements for longer than they want or need to.

Portugal is rightly touted as a global leader for its health-based harm-reduction approach to drug addiction, of which the controversial issue of decriminalisation formed a central plank. It's been two decades since the country's radical reforms came into effect and the achievements are undeniable: drug use as well as drug-related deaths are below the EU average and the proportion of prisoners sentenced for drugs has fallen from 40 per cent to 15 per cent, according to UK-based drug policy foundation, Transform. A central prong of the Portuguese drug policy, alongside decriminalisation, was the expansion of treatment services. While in Scotland they were cut until death figures provoked public outcry. We have no safe-consumption spaces offering supervision, clean needles and antidotes in the event of overdose. We have no drug-testing facilities for users to monitor the potency or potential toxicity of substances. And while Lord Advocate Dorothy Bain announced in 2021 that possession of Class A drugs would no longer lead automatically to prosecution (a move rightly hailed by most as a step in the right direction), drug possession is not always the catalyst for prosecution – rather it's the crimes people commit to obtain money for drugs, or behaviour while intoxicated, that land them in the cells. In the absence of more radical action, the drug sector is simply too reliant on OSTs like methadone (which isn't even the best OST available) and without adequate service provision, it becomes not only a lightning rod in the drug debate but a dead end for many people on it.

Rowdy Yates MBE is an honorary research fellow at the faculty of social sciences at University of Stirling and the honorary vice president of the European Federation of Therapeutic Communities. As well as being eminently qualified, he also has lived experience of arresting the condition of addiction and has been sober for many years. In a briefing paper he prepared for members of the Scottish National Party, in the wake of the publication of the drug-death figures for 2019, Yates conveyed the predominance of opiate replacement therapy in the drug sector:

> 'Opiate-substitute-therapy remains the dominant approach and as a result, other approaches have often been marginalised and squeezed. This may not be intentional, but it is inevitable; like the large, out-of-town supermarket which may not intend to close your village shop but will do so just by being there. More worrying is the fact that two decades of long-term substitute prescribing as the first choice treatment has resulted in an addiction workforce – treatment commissioners, service providers and even their clients – often profoundly sceptical of the possibility of recovery. In many community-based drug treatment services, staff will have rarely seen a recovered addict and often will even discount the possibility. Consequently, we now have a highly medicalised approach to drug use which largely fails to recognise the social and psychological aspects of the disorder and seeks rather to manage the symptoms and reduce the consequent harms.'

Harm reduction should of course be central to any nation's strategy to reduce drug-related deaths but in Scotland, the approach has not developed a great deal. On the other side of the ravine, the recovery movement, where scepticism of harm-reduction strategies is informed by the lived experience of being failed by the various services which deploy them, stands diametrically opposed to the Scottish government's drug sector status quo.

In harm reduction, treatments are informed by a conception of addiction which sits on a scale of relative severity; heroin dependency is replaced with methadone dependency as this is seen rightly as being an incremental improvement. Harm reductionists believe total abstinence is an unrealistic goal for many addicts and that encouraging someone to come off drugs completely may increase the chances of failure and death. There is, of course, evidence to support this position but those who make the argument often fail to add the caveat that it is not reducing or withdrawing an addictive substance which increases the likelihood of relapse or death but the lack of action on systemic barriers in housing, welfare, employment and criminal justice, as well as poor harm-reduction services as outlined earlier, that increase the risk of tragedy.

Meanwhile, in the recovery community there is a widely held belief, again informed by lived experience, that complete abstinence is possible for everyone. That assuming recovery is beyond some addicts represents the soft bigotry of low expectations – a belief that only those with no direct experience of seeing people succeed or of overcoming addiction themselves could hold. And that where people do fail to achieve sobriety, it is often as much to do with the poor quality of information and treatment they receive from harm-reduction services, as well as scant access to them, which impairs their chances of getting clean.

These areas of disagreement are evidence of an experiential gap – a problem of proximity where harm reductionists regard many in the recovery movement as being distant from the evidence while they themselves are regarded by the recovery movement as being remote from the truth of addiction as it is experienced. Harm reductionists are preoccupied with the wicked problems of addiction; the recovery community believes it has found the solution. Currently, however, harm reduction is hegemonic and the radical reforms required to overhaul the drug sector's 40-year-old plumbing is a nightmarish proposition – it prevails because nobody within it can see a way out.

Complicating these increasingly strained dynamics are additional proximity gaps in culture, language and power – themes we will explore in greater depth later. Those embedded in the government's status quo often speak in the dispassionate language of officialdom and conform to a manner of communication which they believe objective and appropriate, but which is interpreted by many with lived experience as arse-covering waffle; this is a temperament which is only possible because officials (who tend to be of a higher social caste) are not directly impacted by the crisis. Meanwhile, those with lived experience, the majority hailing from poorer backgrounds, speak with little affectation and often from the heart, through a thick fog of permanent grief – a communication style regarded by many officials as aggressive, inappropriate and deeply unscientific.

When those in charge are not dealing with the grief of losing loved ones and those who feel they are being pacified and dismissed by them are burying their friends and family multiple times a year, acrimony is sure to develop. Yet despite recent evidence suggesting an addict has as much chance of getting sober through 12-step, abstinence-based fellowships (recovery) as they do with entourages of expensive, professional wrap-around support (harm reduction), the latter model continues to prevail when there is clearly room for more treatment diversity. This is where the complex overlay of class dynamics, institutional resentments and professional jealousies becomes most pronounced – an addict often has as much chance of getting well by following the simple suggestions of other addicts from lower-class backgrounds as they do by placing themselves in the care of highly qualified middle-class professionals.

The drug crisis is a cocktail containing three lethal ingredients. The first is the cross-institutional misunderstanding of addiction. Systems and processes with which chaotic addicts must engage to get support and treatment are often unfit for their purpose. On occasion, they actually enflame already heated circumstances and, as documented in this chapter, can create unnecessary crises or even lead to death. From social landlords evicting addicts (or forcing them to choose between rehab or homelessness), thus throwing their lives into further disarray, to the cycles of incarceration caused by policing and criminal justice systems which do not possess the specialist knowledge (and to be fair, the resources or legal tools) to do much but arrest and prosecute, to the drug sector itself – where services (usually operating 9am to 5pm) and access to treatment (usually made available weeks or months after it is sought), provision is hideously misaligned with the addict's unmet needs.

The second ingredient is the massive class disparity (and the proximity problems and conflicts of interest arising from it) between the people who are predominantly affected by the drug crisis and those who hold the power to do something about it. This creates needless friction, misunderstanding and distrust on both sides of the ravine, where groups who desire the same thing (a reduction in drug deaths) draw wide-of-the-mark conclusions as to the other side's intentions. This class disparity is expressed not simply in the power imbalances (which often exclude those voices the status quo is most challenged by), but in the speech styles, emotional attitudes and the varying senses of urgency and priority.

But the third and often the least remarked upon ingredient is that there is no real accountability for failure. This is best illustrated by the fact those seeking support for drug problems do not always have the same say over their treatment as people with health problems not related to addiction can expect. Their progress along a treatment pathway is often contingent on them demonstrating certain competencies and being seen as deserving or 'ready' – this would not happen to someone with a mental health condition or degenerative brain condition. If addicts relapse, miss appointments or behave in manners judged as unbecoming by drug-sector staff, they can be punished or find their treatment withdrawn – this would not happen to someone with cancer, even if the cancer was as a direct result of an addiction, like smoking.

This lack of autonomy is exacerbated when those presenting with substance issues lack knowledge and the communication skills to self-advocate. They may agree to a treatment plan out of desperation – one which is offered not because it is best suited but because it is all that is available – and then discover it's not what they wanted or what they thought it was. If they relapse because of this, no mechanism exists to ascertain whether the type or quality of treatment they were offered was a proximate cause. Furthermore, if they are mistreated, abused, let down or even die as a result of systemic failure or human error, few processes exist to identify and rectify that – the 'no blame' culture which pervades the drug sector is curious in that no such culture exists in any other area of public life.

Many in the drug sector understand perfectly well the issues I outline here. They are, however, so enmeshed with the Scottish government that they have inadvertently become accomplices in this crisis. In 2021, in response to the lack of accountability I describe, a draft Bill was published for consultation which aimed to 'provide a statutory right to addiction and recovery treatment services' enshrined in Scots law. It was written by Stephen Wishart, whose background is in housing and homelessness, where similar systemic problems exist, and Annemarie Ward, chief executive of campaigning charity FAVOR (Faces and Voices of Recovery) UK. With this rights-based approach to treatment, legal professionals and advocates would have clear instruction on what their clients are legally entitled to, creating vital chains of accountability rather than rehashed, reheated 'treatment standards guidance' – a laundry list of things services agree they will aspire to do which is not legally binding – copied and pasted every few years.

What must be understood is that in the face of this potentially radical reform, many of the drug sector's leading lights said little. Indeed, there is near-uniform silence from certain quangos and charities whenever controversial matters of any kind are contested publicly in a manner which pressurises the Scottish government. When the former minister responsible for drug policy Joe FitzPatrick was forced to resign following the publication of the 2020 figures, the drug sector declined to comment. But when a new ministerial position was created, dedicated solely to drug policy, and £250 million of new funding was announced by the First Minister (as a result of the work of campaigners), the drug sector gushed. The network of organisations regarded as legitimate (which are also in receipt of government funds as well as accustomed to their prominent place in debate and policy formation) tend to remain quiet until the more seasoned lived experience campaigners force change from the outside, through acts of organised campaigning, often at great risk to themselves. Only then, once that dirty work has been performed by those willing to speak truth to power, do the leading lights once again re-emerge, at a politically safer moment, to welcome the wonderful change that has occurred – and re-attach themselves to the new political paradigm they did almost nothing to create.

Back on the street corner, one man is bloated and jaundiced, his black, sunken eyes disappearing into a weathered, expressionless face. I ask him to explain, for the benefit of anyone in doubt as to heroin's attraction, what he feels he gets out of it. 'To explain it, it's as if you're glorifying it,' he says, 'it takes all your woes and your worries away.' What is he trying to escape from? 'Life,' he says, 'everyday life.' What is 'everyday life' for him? He pauses. It's been a while since he was asked a serious question. One which it is assumed he may be capable of answering. 'Wishing you'd never touched drugs in the first place,' he replies.

The others nod knowingly, as do I, before they break into a discussion about addiction, in a manner suggesting they are all, like much of the sector currently failing them, resigned to its permanence. They blame themselves when the fault is only partly theirs. Whatever their shortcomings, they have also been badly let down by a system remote from their problems.

7

Health

The links between social deprivation and health problems are well understood. Yet, as with most action required to truly tackle inequalities, meaningful solutions are a hard sell politically because they involve not simply challenging widely held preconceptions of poverty among the public but also a vast recalibration of society itself. One which those adjusted to their entrenched privileges would regard as an attack. We've known for the best part of 40 years that rapid deindustrialisation was a public health time-bomb – we chose not to act.

In the 1980s, the publication of the infamous Black Report – commissioned in 1977 by the previous Labour government to investigate health inequalities – found significant differences in health outcomes between the social classes – in both sexes and all age groups. The report's introduction stated: 'Social class differences and mortality begin at birth. In 1971, neonatal death rates (deaths within the first month of life) were twice as high for the children of low skilled workers as they were for the professional classes.'

The Thatcher government did all it could to suppress this report's findings and Sir Douglas Black's recommendations – that social inequalities around death and ill-health had to be confronted seriously by raising child benefit, improving housing, and agreeing minimum working conditions with unions – were dismissed before being kicked into the political long grass. Much like the Scottish government's attitude to the many warnings it received with respect to a potential rise in drug deaths, the Conservative government acted against advice about the long-term impact of growing health inequalities.

The Scottish and UK governments may claim they never knew any better. That they did their best. But these are bare-faced lies. The trends already set in motion when the Black Report was published accelerated and became simply undeniable during the pandemic, demonstrated by the disproportionate impact of Covid on working people and the poor.

Back in Possil, the impact of this health gap is both ubiquitous and imperceptible, so adjusted have its residents become to falling ill and seeing loved ones taken before their time, while public officials scratch their heads, puzzled as to why these people can't just get it together.

As dusk falls at around 5pm, the two local pubs are already busy. The demand for alcohol is as abundant as the supply, evidenced by the various outlets on the high street where it's available. Booze, however, is not quite as coveted as a warm, dry place filled with friends to enjoy it with.

'Hello, we're here for the documentary filming,' I proclaim with false confidence as I enter the Saracen Bar, concealing any anxiety that might betray the fact I am not from around here. The woman behind the bar zeroes in on me with 20-yard stare that could leave an impression on steel.

'I don't know anything about it,' she replies with a practised lack of concern.

'What's your name?' I ask.

'I'm the owner,' comes the bold reply. In the time it takes to get to the bottom of the confusion, the pub breaks into a jovial rabble.

An elderly gentleman moves glacially towards the door, in almost comedic fashion, wishing not to draw attention to his sleekit departure. He is immediately spotted and accosted by friends, jovially accusing him of trying to escape the glare of the camera so that he may avoid being exposed as a benefit fraudster. He is slow and unsure on his feet but his friends have keenly assessed that a gentle ribbing is one thing to which he is definitely entitled.

At the sort of wooden tables that always seem to wobble no matter how flat the ground or how sturdy the legs sit elderly men, sipping whisky and half-pint combos. Some are keener to talk than others but it isn't too long before it becomes clear that their cheeky halves aren't all they spend their weekday afternoons nursing. Every person I speak to is managing at least three concurrent and chronic health conditions – known as multimorbidity in the medical profession.

Previously associated with the elderly, multimorbidity is beginning to impact people at a much younger age. In the pub, half of the patrons enjoying an afternoon tipple are afflicted by mobility issues. Two have trouble speaking due to respiratory problems. One has a degenerative brain disorder and struggles to stay in the conversation, while his long-suffering son (and designated driver) brings him another half. It's unwise to speculate as to anyone's age, given the tendency for sick people to appear much older than they are. This is especially so in post-industrial communities like Possil, where ill-health and premature death are par for the course.

One man, Andy, tells me about his first health scare at the age of 29. 'I was in the army when I had my first heart attack,' he says. Of all the gentlemen in the pub he is the widest-eyed, with a radiant skin-tone contrasting so starkly with the paleness of those around him that I wonder whether he is part Italian. Despite his healthy glow (and willingness to share his views on the social history of the area, which he believes has been run into the ground by politicians), he has suffered four subsequent heart attacks since the first. Once a keen marathon runner, Andy is now restricted in his lifestyle and stands out today because he is the only person in the pub who is not drinking. He is here to spend time with his friend, who sits adjacent, panting, his crutches rested against the table, sporting the haunted expression of a man who knows his next breath may be his last; Andy's company, I'm positive, is most welcome. He shares a worrying (and hilarious) story about his fourth heart attack, which he took in his home around the corner from the pub. Given his relative experience of cardiac arrests and his disinterest in going to casualty – where he was convinced doctors wouldn't believe him – he decided simply to wait it out in his home until morning, before going to see a GP whom he trusts.

That GP was Dr Lynsay Crawford, who has worked in areas like this for most of her career. I accompanied Lynsay on some house-calls around the community, keen to get her professional perspective, but found myself struck by how she spoke of her former patients much like you would your own family or friends. 'There's a fatalistic attitude,' she said. She described an interaction with a patient who was weighing up the pros and cons of making a dramatic lifestyle choice to prevent an early death. 'I remember chatting and they said,

"My life's rubbish, I know if I did all these things, I'd live longer but I really don't want to live a longer rubbish life,"' she laughed. 'You've got a fair point my friend, my work here is done,' she joked. 'Another therapeutic success!'

This levity, which arose from the close proximity Lynsay had to her patients, contained a healing quality of its own. She took the time to get to know them, their conditions and idiosyncrasies so well that cracking a joke, or shooting from the hip when someone requires some tough love, was a matter of instinct. I was deeply moved by her devotion to this resilient but challenged community – and how loved she was by those under her care.

Sadly, Lynsay no longer works at the local practice. She was just visiting old friends. Due to the work-related stress and the effect on her home life, she moved on. After years of trying to communicate the impact of poor healthcare provision on both patients and the local practice, she became resigned to the fact nothing would change – much like her patients in the face of their ailments.

'I've been in the papers trying to highlight some of the issues that we had here,' she told me. She even wrote to the health board on numerous occasions but it came to nothing. 'I reached the point where ultimately, I resigned from the practice because I could see that nothing was going to change but my health was going to suffer, and my family life.'

What chance do communities like Possil have when resources are so scant that even doctors who love the areas in which they work, and understand them intimately, feel they have no choice but to practise medicine elsewhere or risk their own health and wellbeing?

'We can't get doctors to come and work in deprived areas,' she explained, 'so if you're living here, you might have to wait weeks to see a GP – who's got ten minutes to see you.' Lynsay identifies funding as the central issue. Specifically, the disproportionate levels of funding granted to affluent communities, where health problems are fewer. 'There was a study a couple of years ago that showed there was about £10 less per patient than if we were in Bearsden [one of Glasgow's most affluent areas] just a few miles up the road. The majority of resources seem to be going to the more affluent areas and not the deprived ones where the health needs are and that has a huge knock-on effect.'

Indeed it does. It's in these conditions that some people simply stop contacting the doctor's surgery and instead self-medicate at one of the numerous local pharmacies, where codeine-infused, highly addictive painkillers, sleeping medications and potent cough mixtures are advertised prominently in the windows and are readily available without prescription – privatisation by the back door. It is also in these conditions that toothaches become abscesses requiring emergency surgery; aches and pains become chronic mobility issues; strange lumps, put quickly out of mind, become stage four cancers and mental health problems and addictions become drug deaths and suicides.

Like many in the face of their health problems, GPs with proximity to communities like Possil, and the people who live in them all over Britain, are fighting a losing battle against distant officialdoms, while being forced to compete among themselves for what scarce resources are made available. 'The disconnect is definitely with those who make policy,' Lynsay tells me. 'The UK and Scottish government, and also the NHS as well as some of the institutions who represent doctors – the BMA and the Royal College of GPs – they do not listen! They don't listen to people experiencing poverty and its effects, and they don't listen to the people working in these communities – I don't just mean GPs but all the other people on the ground seeing the impact of decisions made higher up the food chain.'

Lynsay believes the NHS alone cannot resolve health inequalities because they are caused by wider social issues, like education, unemployment and housing, but that the lack of funding to the NHS over the last decade has exacerbated many of the long-term issues, leaving different factions within the NHS fighting over the same dwindling pot of money.

'All GPs are busy and struggling with workload, so if you decide to move more of the money from the communal pot to fund more deprived practices, then you will be removing money from the more affluent practices and potentially destabilising the provision of care there.' It's this 'communal pot' that Lynsay regards as the problem; when wealthier areas are accustomed to health provision which exceeds the quality seen in poorer areas, and poorer citizens require better care but have grown resigned to threadbare services, the wealthier areas simply dominate the battle for money as nobody wants to put their noses out of joint.

'We proved how much our workload was and that we were still not managing to cope with all the unmet needs of the community. We approached Greater Glasgow and Clyde Health Board for help and stated our case. They spent three months investigating our practice – I suspect hoping to prove we were inefficient or inept – and found we were not. They commented on the quality of care we provided but offered us no extra funding and instead suggested we reduced our standards of care! I went mental at that and in my reply to them asked if they would suggest that to a GP in an affluent area, with vocal and informed patients, or were they just suggesting it because the people of Possil were unlikely to complain or were somehow "lesser"?'

Despite people from poorer backgrounds requiring more health-care than those in more affluent communities, they receive less. In particular, less time with their GP, resulting in poorer quality encounters. Wealthier patients' health problems tend to be less complex – multimorbidity and poverty are synonymous – requiring less time to explain during a consultation, yet they may receive, on average, a minute or so longer with their physicians than someone from a poorer area. Analysis from the Health Foundation in 2020 found that patients in the richest areas receive 11.2 minutes on average while those in the most deprived communities get 10.7 minutes. This is not an anomaly. This imbalance is written into health systems and is known in the field as the 'inverse care law'.

First proposed by Julian Tudor Hart in 1971, the inverse care law is the principle that the availability of quality medical or social care often varies inversely with the needs of the population served. A pun on the inverse-square law, a term and concept originating in the field of physics, it attempts to draw out a paradox which is central not only to health inequality, but all inequality. In the paper, published in the *Lancet*, where he first proposed the inverse care law, Hart wrote: 'The availability of good medical care tends to vary inversely with the need for it in the population served. This operates more completely where medical care is most exposed to market forces, and less so where such exposure is reduced.' He later clarified his thesis, adding: 'To the extent that health care becomes a commodity it becomes distributed just like champagne. That is rich people get lots of it. Poor people don't get any of it.'

Back in the Saracen Bar, the drink is flowing as I make my way to the lounge, where the karaoke is being held. It's not as busy here, though less awkward with music to fill the silence. A rotund gentleman approaches to humbly inform me that he is the best singer in Possil – a likely story. Though, given that nobody so much as rolls their eyes let alone rises to test the veracity of his claim, perhaps there's some truth to it. There is not a glint of bravado in his eye, only pride. He takes the microphone in his hand and I am almost moved to tears, not just by the beauty of his voice but the poignancy that it should bellow so gracefully from his lungs, despite evident physical challenges – his throat seems obstructed and he gasps for breath as he belts out his tune. Far less talented vocalists enjoy long careers after receiving professional training at performing arts schools. Yet, here he is, in the twilight of his life – which here is around the age of 60 – singing purely for the joy of it and hitting every note with a tenderness not often associated with these parts. He receives a restrained applause from his fellows, who haven't yet imbibed the volume of alcohol required to publicly display emotion, before walking back to the table to offer me a drink, which I politely decline. I explain I no longer partake due to alcoholism. He nods knowingly, requiring no additional explanation.

'I wish I could stop, but what's the point?' he sighs. 'I'll be dead soon.' He exhibits a specific brand of fatalism that is hard to judge or discourage. When people of a certain age speak, they convey a wisdom irrespective of their intent. Encoded in his resignation is everything he has learned from the battles he has fought and lost throughout his life. There comes a point when it becomes easier, rational even, to write off the losses and simply live your life, whatever folly it may entail. The years he may conceivably add to his life by a sudden cessation in alcohol and tobacco he will likely lose dealing with the stress and isolation that often follow kicking a habit.

Perhaps the saddest aspect of life here is that the places many people associate with happiness, pleasure and connectedness also play a decisive role in their needlessly abbreviated lives.

At various stages in our scattered conversation, the singer stops to thank me for 'coming here'. I am touched, though slightly confused, by his strong sense of gratitude. I fear he has mistaken the cameras as evidence of our elevated social status, bowled over that we would burden ourselves by abseiling into his community from our ivory tower at the BBC.

Feeling the strain of alcohol flowing around me, I become restless and go outside for a cigarette. In the doorway sits a man in a wheel-chair, puffing a smoke, looking across the high street to an overgrown public square. Building rubble, abandoned after a fire, is encircled by hideous barbed-wire fencing. I step into the middle of the street and turn 360 degrees, curious to see if it is possible to gaze in any direction without being confronted by evidence of dereliction, but its demoralising presence is an all-encompassing fact from which I cannot avert my eyes. The streets are littered with discarded food packaging and chewing gum. Even some of the local businesses thriving here have thrown in the towel when it comes to the general upkeep of the community. Rubbish bags are dumped on disused land and overflowing bins line parts of the street, creating an impression that tending to one's refuse would be a thankless and futile task.

The litter problem is often attributed to the locals, regarded by their betters as irresponsible and lazy. You can infer this from the posters plastered around the community, reminding local people that the street is 'not your personal bin'. Often these posters line the periphery of abandoned demolition sites and large plots of vacant land that have lain barren for as long as 20 years. One could argue that the same local authority that prints and distributes these posters is the sole arbiter of the dereliction and resulting mess. If there is one factor shaping the alleged low aspiration of the community, it is surely the fact nobody can look out of a window, leave their home or walk five minutes down the street without encountering a vast strip of ugly, disused real estate for which the local authority is responsible.

Despite this being a small area, many locals feel dislocated from their surroundings. There is great confusion within the community about who owns the land and who is responsible for its upkeep. This creates prime conditions for fly-tipping. The council encourages people to report littering but can't get its own house in order when it comes to removing discarded furniture and household items from the pavements. It can take as long as two weeks to get a pickup, forcing many to discard it by the roadside, which simply accentuates the sense of lawlessness.

Derelict and disused land, to the extent that it exists in Possil, creates an aesthetic stress, which, along with the littering and the sense of resignation, contributes significantly to mental health problems.

As I take the last draw of my cigarette outside the pub, the pale, breathless gentleman accompanying Andy struggles out on his crutches. I ask him if he needs a hand to his taxi. 'No, son,' he recoils with misplaced pride, before adding: 'This is what happens when you have a stroke.'

This is a community defined culturally by its poverty-induced health and social problems. Possil tends to only make the news when morbidity statistics are published or when someone gets murdered. We rarely hear about the local 'Dancing Grannies' group that meets every Monday, offering light exercise, fellowship and a great selection of biscuits to women experiencing loneliness in old age. Few column inches have been devoted to the various grassroots organisations in the area that teach local people how to grow their own fruit and vegetables or the environmental activists that campaigned successfully to have a 6km path laid down through an amorphous swathe of abandoned land, transforming it into a sprawling public space, teaming with wild- and plant-life. In fact, the only time you will hear about the beautifully serene canal that runs through the heart of this community is when a body is being fished out of it.

Sadly, as is usually the case, the statistics fail to capture the real story here. Indeed, statistics reported with little context place an additional wedge between the community and those who observe it from a distance. When health stats are published, we only get a sense of those who are afflicted by illness. Those who die prematurely. Even then, there is a feeling of inevitability about these deaths. What we get no sense of is the trail of bereavement and the shadow of grief that looms over everyone who lives here. For every heart attack, stroke, drug death, murder and suicide there are countless family and friends on the periphery who must absorb the loss of a loved one while trudging on with their own challenging lives. Here, everyone is mourning a loss of some kind. Some even mourn the old tenements, demolished long ago – the land on which these 'vertical villages' once stood remains barren, disused and overgrown. The lifestyle so often associated with people who live in post-industrial communities like this – overeating, excessive drink and drug use, gambling – provides temporary comfort and fleeting familiarity and continuity in what can be a chaotic, deeply disheartening and tough existence. Many adopt fatalistic attitudes towards their own health because it's a way of exerting control over external and internal circumstances they understandably feel are beyond them.

Back inside, a younger man approaches and welcomes me to the community. He is the youngest patron in the pub yet seems strangely aged by booze. 'Everybody knows me,' he claims – another way of saying he feels connected socially. His sense of connection may also be why he is so welcoming. Often, it's those who feel little connection who become insecure and hostile to those with whom they are unfamiliar, whether they be strangers, immigrants or media wankers like me.

He asks me for a cigarette and we head back outside. As we chat about the documentary, our conversation is interrupted by a phone call. It's a young woman looking for the painkiller tramadol, which he is prescribed by his GP but no longer uses. Instead, he sells the tablets to make a bit of money. After the call, we bid farewell as he heads into the city centre to close the deal. The other pub, across the road, is now teeming with custom. A poster, emblazoned on the blacked-out window, reads: 'Come in for one of your five-a-day.' The sign is certainly poking fun, though I am not quite sure of the target audience.

According to the Scottish Multiple Deprivations Index, women here live an average of seven years longer than men – life expectancy for men is 66 while female life expectancy is 73 – though both sexes have the lowest life expectancy of all neighbourhoods in this city. A relatively high percentage of people are also limited 'a lot' or 'a little' by a disability, which is evident the moment you hit the high street. The number of people hobbling around on walking sticks, frames, wheelchairs or mobility scooters is astounding. The neighbourhood has a lower-than-average employment rate and a high number of young people are in neither education, employment nor training. The number of children living in poverty is considerably higher than the city's already shameful average.

This community was literally built around industry. The industry is what gave this community its centre. The products that were designed and manufactured here can be found on every continent. In other more affluent parts of the city, the industrial history of this community is not just celebrated but mythologised – its hollowing out, of course, omitted because it undermines a sense of middle-class nostalgia and the comforting myth of progress. For half a century, attempts at regeneration have failed to account for the mass severing of social bonds and connections that occurs when families and individuals become displaced and dislocated by the very 'regeneration' sold to them as progress. For those who remain, it's often not by choice.

Attention is often drawn to the millions poured into such communities for housing, which is certainly welcome but not enough. What this community has lost cannot be replicated with a shiny new initiative. This community has suffered a deep spiritual injury, leaving successive generations socially and economically rudderless and more disorientated than the last. There is no longer a sense of place or history in which people can orientate themselves – only the grim narrative of disrepair and political neglect. Most green space that exists here is a result of nature making its incursion into the disused land, bursting from the rubble abandoned by construction companies, which the council seem unwilling to pursue with the same vigour as they do litterers. Trees and plant-life spring from the social graveyards of demolished schools and tenement communities to the extent that this area has become a globally renowned conservation site for wild birds – something most of the local people are not aware of.

In a community where 100 per cent of residents live within 500 feet of a derelict site, and where the local leisure centre was recently demolished, local pubs compete with a solitary, soon-to-be-condemned community centre, using what little public space remains for people to mingle and be. This is not simply a crisis of poverty, but a far more fundamental one of human social connection.

After my trip to Possil, I contacted Dr Crawford again, to hear more about her theory that poorer communities were being deliberately deprived of resources to maintain higher standards in favour of wealthier ones, where health problems are less prevalent and complex. 'I firmly believe that resource should go where it is most needed,' she told me, 'but I suspect that the decision makers have an eye to who votes for them or pays their wages.' You can see why her patients loved her – she really did care about them. Crawford questions the wisdom of splitting resources equally among all practices regardless of need and believes this funding formula is why the gulf between social classes with respect to health is widening. 'The overwhelming body of evidence shows those in poorer communities have worse health (at a younger age too) and have a greater need for funding to reduce health inequalities. It is clear that equal distribution of funding is leading to greater inequality.'

In areas like Possil, across the UK, people in most need of care are simultaneously the most distant from the health services that could improve or save their lives. This low proximity is mirrored by decision makers at local and national level, remote from those on the ground who understand better the complexities Britain's poorest communities face. Meanwhile, politicians in charge of the purse strings, keen to remain in the good graces of the middle classes, are far too attuned to their own short-term electoral interests, at the expense of longer-term health equality. And every time they act in those interests, thus postponing confrontation with the reality that wealthier citizens hoard too many resources, poorer people fall ill and die. What should worry all of us is just how adjusted those in most need have become to this dangerously imbalanced health status quo and how this low expectation lets politicians off the hook. But perhaps the most shocking aspect in the ongoing public health debate, so often centred on notions of individual responsibility rather than systems which demonstrably favour wealthier citizens, is how distant most people remain from the ugly, irrefutable truth: poor health is absolutely a choice – a political one.

8

Welfare Reform

Britain's grim inequalities are historic. While many do desire change, reforming institutions which were designed for the twentieth century is undeniably hard. It's easy to assume that a lack of progress is simply down to our best efforts being hampered by a century of institutional baggage that requires time to sort through. And yet, while the pace of reform in some areas has been slow, in others it has occurred at breakneck speed. What these more recent reforms, particularly in welfare, reveal is that the governments can act swiftly if they want to, and that the growing sense of embattlement and crisis many of Britain's poorest and most vulnerable experience is not just down to historic factors difficult to unpick – their nightmares are induced methodically by the state.

At the dawn of 2013, only a few days sober and feeling a little fragile, I received a brown envelope from the Department of Work and Pensions. Letters from the Job Centre were never pleasant to read but, in my experience, they were more confusing than distressing, reading impersonally, as if generated by a computer.

I was used the discomfort that came with dealing with the DWP but not even a stiff drink could have prepared me for the letter I received that frosty January morning as I unsheathed it from the envelope and read, to my horror: 'Amount owed: £3,998.35. We are writing about money you need to pay back.' According to the letter, I had been overpaid nearly £4,000 in Income Support. The letter included no evidence of this figure, why I had been overpaid or how the figure had been calculated. All it told me was that I owed the money and that I should pay it all back immediately. The letter was written in a very stern and unsympathetic tone, one that I suspect was designed to illicit keenness to comply at the earliest convenience. But I was an alcoholic just days into my recovery, having spent the previous 48 hours tossing and turning on the couch, sweating, falling in and out of night terrors and sleep paralysis, going through a difficult and painful withdrawal.

The letter sent me into a blind panic. It felt like I was being accused of a crime. Fortunately for me, someone in my recovery group at the time was also a welfare rights officer. After hearing me share my anxiety around the issue at a meeting, he offered to help me. Within three weeks, the case was dropped. As it turned out, the DWP had insufficient evidence that I had been overpaid or that it was my mistake. However, had I not been fortunate enough to have come into contact with someone able to advocate on my behalf, I would have been compelled to pay the money back regardless.

A few weeks later, a family member went missing after contacting me directly to tell me they were going to take their own life. After hours of frantic searching by police, including a police helicopter, they were thankfully tracked down and tragedy was averted. Due to this, I was unable to attend a Job Centre meeting, but when I informed them, I was told that I would be 'sanctioned' for six weeks – 'sanctioned' being a term I had never heard before in all my years of dealing with the benefits system. It meant, of course, that my benefits would not be paid for a month and a half. I went in to speak to them directly and thankfully the staff member was not only sympathetic but able to exercise some personal discretion and reinstate my benefit. But even that was, I sensed, contingent on my ability to communicate in a specific manner. To express clearly what had occurred. To convey my reasonableness.

Of course, to my left and right, in a busy and stressful open-plan office with a security guard on the door, other claimants, perhaps lacking my communication skills and acquired sense of officialdom, did not fare as well. I could hear them being chastised, in full view of everyone. It reminded me of school and of being in police custody. I left, relieved to be out of there, and thought little of it, and within months enrolled in a journalism course at a local college. Unbeknownst to me, I had dodged a rather large bullet, as that year, evidence would begin to emerge that what I had experienced was part of a broader trend. One where benefit claimants would gradually become subject to a tougher regime of welfare conditionality brought about by the coalition government's now infamous welfare reforms.

Department of Work and Pensions statistics published in 2013 revealed that 553,000 sanctions were handed down between November 2012 (when the new regime started) and June 2013. In November of that year, the *Guardian* reported on then-Work Minister Esther McVey's assurance to the British public that these steps had been taken against people who had missed appointments because they were deliberately avoiding having to get a job. She said: 'Sanctions are used as a deterrent. The people who get sanctions are wilfully rejecting support for no good reason and if there were a reason, there is something known as "good cause", so if that seemed true and genuine, you'd have good cause there to not have a sanction, plus there is a process in place just to ensure we are getting it right.'

As befogged as I undoubtedly was in the early days of my first serious attempt at sobriety, I understood, on some level, what lay at the root of the frightening letter and the unreasonable sanction. This was a public institution's attempt to save some money by making the experience of claiming or being on benefits so miserable that I would take the first job going simply to be rid of it all. I did not believe the DWP had a duty to make it a pleasant experience; the level of benefit and the way it's accessed clearly must be carefully considered to deter abuse of the system. Most people understand this. But I was also emerging from a serious alcohol and substance misuse problem with a medical record thick with hospitalisations related to it. Rather than incentivise me into work, the DWP's strategy filled me with fear, dread and shame.

In my dealings with them, particularly written, I came to understand that whoever thought these stern, cold communications would be useful in encouraging people off benefits and into work was operating from assumptions rooted in a different emotional universe from my own. Encoded in these letters were certain false beliefs about how human beings respond to provocation, humiliation and aggression. What I concluded, after much reflection, was that the government's conditionality regime was deeply, sometimes fatally, emotionally illiterate – if you want to persuade a vulnerable person, particularly someone with a history of trauma and in recovery from serious addiction, to do something, you must take a more sophisticated approach than a local thug. Of course, if you had never suffered the levels of deprivation and adversity experienced by many who end up on benefits then this may not occur to you.

The true scale of dysfunction brought about by the financial crash and the welfare reform which followed began to dawn on me years later when I attended an AGM at Kirkcaldy food bank (KFB) in Fife. The small church hall was packed to capacity, with additional seating brought in to meet the demand. The fact the event was so well attended took me by surprise – AGMs can be tedious affairs – but that wasn't nearly as surprising as the main item on the evening's agenda. Aside from the usual business of accounting for income and expenditure, and presenting members with an annual report detailing the sharp rise in food-bank use as well as data showing specific reasons people were accessing the service, the committee also had to hold a vote to alter a tenet of their constitution – when it was written in 2017, they did not foresee the food bank being needed for any more than two years.

Between January and December 2017, KFB distributed 121,311 meals to the local community through its network of five distribution centres. A total of 4,046 children received assistance in the same year. Most individuals and families accessing the food bank were self-referred. When they presented at the food bank, they were asked to detail the nature of the 'crisis' they were experiencing. Benefit sanctions, benefit changes and benefit delays accounted for 41 per cent of food-bank referrals in Kirkcaldy at that time and was representative of the wider food poverty crisis across the UK.

As we have already seen, food-bank use has soared in recent years across the country and, by some strange coincidence, so too has the roll-out of Universal Credit – the UK government's attempt to harmonise the welfare system, which Labour MP Frank Field correctly described in 2017 as 'an obstacle course of unreliable computer systems, arcane rules, massive delays and maladministration'. Figures from the Trussell Trust showed that between March 2017 and March 2018, 1,332,952 emergency food supplies were distributed to people across the UK.

Nationally, low income is the main reason for referral and accounts for nearly 30 per cent of food-bank use. But, as reflected in Kirkcaldy and every other area of the UK, welfare reform also plays a significant part in driving food-bank referrals. Benefit delays accounted for 24 per cent of referrals while benefit changes were cited by 18 per cent – 42 per cent in total, similar to Kirkcaldy. The correlation between Universal Credit and food-bank use becomes more pronounced when you map the roll-out of Universal Credit directly onto the areas experiencing the sharpest rises in food poverty. This research revealed an average increase of 52 per cent in food-bank usage in the 12 months following the roll-out dates in each area. Whereas food banks outside the Universal Credit roll-out zones experienced an average increase of just 13 per cent.

In 2018, I travelled to the Scottish Borders to learn more about rural poverty. I met a man named Glen who, a few years prior, had become homeless and suicidal following the breakdown of a long-term relationship. People who have resorted to rough sleeping operate on the edge of chaos, though support services often only become available to them once they have become engulfed in a personal crisis, by which point their problems are more complex, severe and costly. The untenably high threshold for what constitutes vulnerability and crisis plays a decisive role in driving the death, despair and dysfunction which has become commonplace.

'I'd been living with my partner and the relationship broke down and she wanted me out,' said Glen, an immaculately dressed, middle-aged lorry driver. 'I'd been made redundant at the same time. The only place I had to sleep was my car.'

'How long did you sleep in your car?' I asked, expecting him to say, 'a couple of nights'.

'Five months,' he replied, with a dead-pan expression and furious gaze. 'I tried to get help with accommodation, tried to get help with benefits and the problem is, each one would say they couldn't help me and they would send me onto the next.'

Glen described the torturous bureaucracy that our most vulnerable citizens must navigate in order to obtain even the most basic support. A bureaucracy – acting as an administrative perimeter, preventing access for many in need as well as insulating those working within services from direct accountability – that is not so much a system but a jigsaw puzzle that works against itself and those who come to depend on it. 'You can't get benefits without an address. Accommodation providers won't give you a place without an income.' He began to well up, the memory clearly still very painful. At the root of the anguish lay the certain knowledge that he had become invisible in the eyes of society: 'I no longer existed,' he told me. 'I had nothing.'

For those of us who have never experienced what it's like to have nothing, such a claim by someone else can leave us sceptical. The notion that, in a society as prosperous as ours, people can still fall on such hard times is not only hard to believe but almost impossible to conceptualise. 'I was walking into offices,' Glen said. 'I was scruffy, no sanitation, I was underweight and I was obviously breaking down. I feel if a stray dog had walked in, they'd have done the right thing by that dog and looked after it.' He welled up again. 'I'm better than a dog.'

It was this realisation that he was no better than an animal in the eyes of many who were responsible for helping him, that he, like John, my late friend from Glasgow, began contemplating suicide: 'The only little bit of control I had left was to choose the last day that I spent on this Earth.' This contemplation soon became an obsession. The heartache from his relationship ending, coupled with the acute and constant strain of social exclusion, and the frustration of dealing with a patchwork of under-resourced services, themselves operating at capacity, left Glen, in his mind, with only one option.

After the usual planning stage which precedes many suicide attempts, Glen's life was thankfully saved by a random police patrol near the remote location he had chosen to take his own life. You'd think, after such an ordeal, that this episode would represent a conclusion to Glen's frantic search for support. But even those who suddenly meet the criteria to be classified as vulnerable will face an uphill battle where public services are concerned. While in hospital, he received no support to access housing services, which left him at risk upon discharge. Eventually, he signed himself out of hospital and was shuttled around bed and breakfasts for a week before moving into a homeless unit. Unfortunately, many in the unit also had alcohol and drug problems, which brought with it the usual dangers – Glen's money and belongings were often stolen. It took a further three months before he was rehoused in the flat he lives in now.

Glen's survival, aside from the good fortune that police were on patrol the night he tried to take his life, was ultimately down to his own resilience as a human being. This is why so many other people don't make it. The adversity, compounded by the dire state of support services, quite simply overwhelms them. The environments they inhabit have become so treacherous, awash with violence and drugs, that periods of personal progress are often undermined by nightmarish social settings or public services operating under increasing strain. For people facing mental health challenges, in the context of family breakdown or even violence and abuse, simply encountering the Welfare State, whether it's actively enforcing compliance or clumsily failing to detect complex personal circumstances, is hellish. But for people struggling with those challenges while also living with disabilities, it is life-threatening.

For a real insight into the toxic social impact of welfare reform, the most useful thing you can do is spend time listening to a person living with a disability describe their experience of applying for benefits to which they are legally entitled. Benefits they need to survive. I've spent time with countless individuals whose lives have been turned upside down by short-sighted and callous policies. But my time with one man, as he relayed his experience of dealing with the DWP, while struggling through the tics associated with his Tourette's, will never leave me.

In 2018, I met with Steven, an artist, at his home in the Scottish Borders. His immaculate dwelling, decorated by his own works, felt more like a religious temple than a home. He showed me around his modest flat and, as I inhaled the potent incense burning, I realised I was missing a trick where domestic comforts were concerned. Every colour, every scent and every texture in Steven's home elicited a calmness. I was there to discuss his recent experiences with the DWP but part of me just wanted to lie down on his couch and drift off to sleep. But every now and then, his Tourette's flared up, causing involuntary physical movements, gestures, facial expressions and verbal outbursts. I had never encountered anyone living with Tourette's syndrome before and I found it unnerving. I understood that the condition manifests varying degrees of severity and it was abundantly clear to me that Steven's case was on the extreme end of the spectrum.

I began by asking him what it's like when the brown envelope falls through the letterbox – an event many people in receipt of benefits dread. Correspondence with the DWP can be an agonising and confusing process. Whether the long and confusing forms, the long waiting times when attempting to contact by telephone, the length of time claimants have to wait for a decision and the often inevitable and protracted appeals process, brown envelopes dropping through the letterbox are feared the nation over because they rarely deliver good news. 'It's gut-wrenching,' he said, 'my experience of going into the benefits system and that's what it is – it's a system – excuse my French but it's an absolute fucking nightmare.

'I was actually waiting weeks and weeks and weeks to find out whether I was getting that form or whether I was going for a onc-to-one assessment,' he told me. 'Going through the assessment and then waiting again months and months to hear about whether you've been awarded the benefit, or they've said "No, you've not got enough points", and then worse than that you have to appeal the thing.'

The appeals procedure is just one horrid aspect of the deeply flawed Personal Independence Payment process. It involves hours of paperwork, filling out forms which would trigger anxiety in even the most mentally resilient people. But for those living with chronic physical and mental health problems, the administrative dimension of the process is not even the hardest part. The assessment aspect, which involves claimants travelling to an interview or having an assessor visit their home, is often humiliatingly intrusive. Analysis of Ministry of Justice data in 2019 showed that disabled people are almost twice as likely to win their disability benefit appeal than they were 10 years ago, which speaks to the problems inherent to the assessment process – there was a time when most initial decisions not to award people living with disabilities the level of benefit to which they are entitled were wrong.

'Like a year ago I had to go for my PIP assessment – that was pretty much a nightmare because what happened was I had a full blown tic attack during my assessment.' A tic attack presents much like a seizure – Tourette's sufferers become overwhelmed physically and must remain stationary wherever they are until the attack passes. 'I fell out of the chair, onto the floor in front of the assessor,' he explained. 'What ended up happening was the assessor was going to postpone my assessment to another day and the woman who was with me, Laura, said you can't do that because it'll be worse for him when he comes back the next time. The assessor was totally shocked, she didn't know that Tourette's syndrome could be like that, she was stunned. She was like, "I've seen enough, I've got enough evidence."'

Steven's assessment is a prime example of another flaw inherent to the assessment process – assessors are not medical professionals. In the assessment process – which looks at whether you are able to perform certain activities, covering both physical and mental health, and awards points for the activities you are not able to do or struggle with due to your health condition or disability – a person with dementia could be asked to recall how their condition affected them in a previous year. An assessor might enter someone's home and ask them for a cup of tea, later citing the fact they were able to make tea against them in the final analysis of their disability. The assessment process can often fail to account for how many mental and physical conditions fluctuate; some days, a person may find themselves able to walk without much pain but in agony and immobile the next. And many people with less visible disabilities are often denied payment due to the inadequate nature of the assessments, which leave too much discretion in the hands of individual assessors with varying degrees of understanding (and bedside manner) regarding disabilities, generally.

'That's the times when I've come so close to checking out here,' Steven confided. 'I'd given up because I was fucking broken, I thought I can't cope with this.' Steven was describing the suicidal ideation triggered in many people when they are compelled to deal with the DWP. A spate of suicides related to welfare reforms has been reported in recent years, including some people even leaving suicide notes referring to problems with the DWP. This is a pattern increasingly seen in those living with disabilities, prompting the government to begin monitoring suicides where they related to benefit claimants.

'I couldn't see any way out of the situation that I was in,' Steven said. 'There are people that I know this year who have taken their own life, really sadly, like very close friends, you know? They couldn't hang on.' He broke down, sobbing uncontrollably, as he recalled the lives lost.

Once Steven felt more composed, we turned back to the assessments process. Did it make him angry? 'Really angry ... people who do it don't know what it's like, what you've gone through on a day-to-day basis. All anybody that's got mental health issues or long-term disablement are asking for is empathy, love and compassion – but you don't experience that within the system, within the benefit system.'

I put it to him that there are people out there who, despite the immense pain he had suffered, would regard him eventually being awarded PIP as evidence that, for all its imperfections, the system did work in his case. The question visibly disturbed him, clearly evoking the painful memories of feeling unseen and of friends lost. I immediately regretted playing devil's advocate. Steven stood up and stumbled to the floor. In a panic, I asked if I could get him anything. 'No, I'll be alright mate,' he assured me, before lying down and proceeding to enter a violent seizure.

Incapacitated, he rattled like a million volts were passing through him yet remained conscious throughout. I sat there stunned, my crew similarly so. We didn't know what to do. My producer wisely fetched a glass of water and set it down beside him, but all we could do was wait like he'd asked us to. For us, this represented the most frightening and shocking experience so far in the process of making our documentary. For Steven, it was just an average day.

'It just explodes right through me,' he explained later. 'I feel like my whole nervous system's just firing, the electricity. It gets hard for me to speak as well, and breathe. It can take me a while to get back up again, it's like I lose all the feeling in my muscles. Sometimes my tic attacks will last for 10 minutes, or maybe longer, maybe an hour or something. The worst was about an hour and 50 minutes.'

As I looked on, feeling quite helpless to intervene, I wept. And then came the anger. If only Iain Duncan Smith could see what I had just seen. If only the average Joe, whose opinion of benefits is shaped largely by those in the media who are rarely qualified or experienced enough to report it, was in the room with me then. Would it be so easy to make baseless claims about 'scrounging'? Would the hyperbole, the prejudice and the bigotry roll so easily off the tongue then? What if they were sat in Steven's armchair, and not their own? From this proximity, where I could reach out and touch this man in a desperate and futile attempt to express concern and compassion, would it be so easy to draw the wrong conclusion as to why he is on benefits?

For countless benefit claimants, the system works reasonably well. For many, welfare reforms have simplified the process of applying for benefits as well as removing certain stigmas associated with so-called 'legacy' benefits – the previous benefits Universal Credit was partly designed to replace – Incapacity Benefit, for example, which became synonymous with long-term unemployment and 'scrounging'. Claimants also have more flexibility in how they are paid; they may withdraw a portion of their benefit early in the form of a small loan, which is then paid back through deductions when the payments begin. Of course, the system tends to work best for those who enjoy more favourable contextual circumstances – the short-term unemployed, people without serious mental health problems or substance misuse issues, people with fixed addresses and those who communicate well.

Public attitudes on matters surrounding welfare, such as reducing overall spending, conditionality (what someone has to do to stay in receipt of benefits, like keep a job diary, attend meetings, etc.) and which groups should be entitled to it and which should be restricted, are not as hardened as the recent welfare reforms may suggest. Most British people believe in the idea of a social safety net for the sick, the elderly and the unemployed. Where opinion varies is in the extent and duration of the support available to specific groups.

Public support for welfare spending generally, however, has been in long-term decline, and while the last five years have seen, at most, a very small reversal of this, overall backing for welfare spending remains much lower than it was historically. NatCen Social Research's British Social Attitudes report in 2015 found that support for increasing taxes and spending more on health, education and social benefits fell from 63 per cent in 2002 to 32 per cent by 2010 – nearly half – and had only increased slightly to 37 per cent by 2014. The level of agreement with spending more on welfare benefits for the poor fell from 61 per cent in 1989 to 27 per cent in 2009, and remained low, at 30 per cent in 2014. Some benefits are more popular than others.

When it comes to extra spending, the public is far more likely to prioritise pensions and benefits for disabled people, and far less likely to prioritise spending on benefits for single parents or unemployed people. Sixty-seven per cent place spending on pensions first or second in their priorities for extra spending on welfare, followed by 60 per cent who chose benefits for disabled people. In contrast, just 13 per cent said benefits for unemployed people should be one of the top two priorities for additional spending.

According to the report, there are two determinants that predict a person's hostility or openness to a generous Welfare State – political beliefs and social background. Conservatives on high incomes with no experience of the benefit system are likelier to be in favour of cutting welfare spending, while Labour voters on lower incomes with some experience of welfare dependency are likelier to oppose welfare cuts. This should not surprise anyone. However, what this does reveal is that the UK government does not necessarily require broad public support for its welfare policies, only the support of its key electoral demographic. In democratic terms, this creates a scenario where the ruling party (in a country where more people usually vote against it than for it) may drive through political reforms which are not only unsupported by a majority of the population but which also harm substantial sections of it. Indeed, the misguided public opinion on welfare reform exists as a direct consequence of a concerted government campaign of misinformation and suppression of statistics relating to its disastrous impact. Misinformation which is more readily accepted if you have little-to-no proximity to the reality of poverty.

What evidence existed to support the UK government's decision to intensify welfare conditionality in 2012? What evidence existed to support the hypothesis that by making the process of applying for and being on benefits so unpleasant that people find it confusing, humiliating and frightening those currently claiming would miraculously transcend their multiple disadvantages? The answer: there was no evidence. All the evidence runs contrary to the policies that the UK government implemented.

The Welfare Conditionality project, running from 2013 to 2018, conducted analysis on both the impact of welfare reforms as well as the practices that underpin them. Concerning homeless people, the report found that benefit sanctions caused 'considerable distress and push some extremely vulnerable people out of the social security safety net altogether' and that 'dealing with the "fallout" from sanctions diverts support workers away from assisting with accommodation and other support needs.' For disabled people, the report found that the Work Capability Assessment – in which people with disabilities must prove they are unfit for certain types of employment – is 'intrusive, insensitively administered and regularly leads to inappropriate outcomes in respect of disabled people's capabilities to undertake, or prepare for, paid employment'.

Where job seekers are concerned, welfare conditionality 'did not prompt behaviour change' and claimants felt there was 'a lack of clarity or warning that their behaviour was sanctionable, that work coaches were too quick to resort to the use of a sanction, and that sanctions were disproportionate to the alleged transgression'. For many lone parents, it found that 'insufficient account is taken of caring responsibilities when claimant commitments are devised' and that many lone parents were sanctioned as a result of 'unreasonable expectations, DWP administrative errors, or failures of comprehension rather than deliberate non-compliance'. Across the board, the persistent threat of sanctions caused 'extreme anxiety, even when not enacted' and had a multiplicative effect on the adversity many people were accessing benefits to manage.

Even some Conservative MPs have spoken out against the harshness of this regime, forcing moderate concessions from the government, such as a reduction in the length of time claimants must wait to receive their first payment. But these concessions will do little in the grand scheme of things because the orthodoxy informing the culture at the DWP remains the same: an assumption that tough social cues and incentives, designed to deter people from benefits by inconveniencing, humiliating or frightening them, will achieve something other than increased adversity and social mayhem.

I don't believe that Conservatives who support these measures are doing it out of hatred for the poor. I think they genuinely believe work is the best route out of poverty, that a less generous Welfare State incentivises people into work and that, for all the criticism they will receive, tough love is the best medicine. This is a sentiment which also resonates on some level with most people and in principle, that's not such a bad thing. But what we now know about the culture of welfare conditionality which has characterised the benefits system for the last decade or so is that rather than getting the poor and vulnerable into work, it's pushing many deeper into the quicksand. And while the public justification for welfare reform remains that it's a cost-cutting exercise, there are arguably other areas the government could target – not least tax avoidance and evasion which dwarfs the benefit bill by several hundred billion.

There remains one aspect of British welfare reform which is dangerously understudied and reported – the insidious influence of the private sector and the behaviourist practices of corporate America.

Unum Provident is one of America's biggest insurance companies; a quick online search will reveal the company – which specialises in disability insurance – is no stranger to lawsuits. One class action lawsuit, often referred to as the Unum/Provident Scandal, arose from an investigation by the Department of Labor into the business affairs and practices of the company, in which it was alleged that Unum (known then as Unum/Provident) denied or terminated thousands of legitimate disability claims, beginning in the 1990s and continuing until 2002. The Department of Labor concluded that the company had acted in an 'unfair and unjust' way by deliberately resorting to fraudulent tactics of denying legitimate claims as a cost-cutting measure.

Under court order, Unum was compelled to reopen over 200,000 denied claims and re-evaluate them based on their merit. To ensure fair and just review and handling of all further policyholders' claims, Unum was charged with overhauling the methods by which they evaluated and processed claims and ordered to pay a fine of $15 million to several states. Despite the financial and reputational ramifications of the scandal, as of 2007, Unum conceded that just 10 per cent of the claims flagged for re-evaluation under the terms of previous legal settlements had been reopened.

Back in the UK, around two years later, in direct response to welfare reforms that included the Work Capability Assessment (WCA) becoming the subject of increasing public and media scepticism, hundreds of disability support groups began springing up around the country. Some offered fellowship and social solidarity. Some advocated on behalf of those living with disabilities and others took the form of pressure groups. But one woman, an army veteran and former health professional, began a period of pioneering research, broad in scope and rich in detail, which, by 2019, would recontextualise UK welfare reforms not simply as cruel and costly, but also as evidentially baseless.

Mo Stewart, a disabled veteran of the WRAF medical branch, has led the research into the intellectual and corporate influences behind the catastrophe of welfare reform. Her Preventable Harm Project(2009–2019) drew from her own independent research as well as disparate pieces of academic research into both the causes and effects of welfare reform. In doing so, Stewart created a picture which contradicts that which the UK government and the DWP have been painting. Her painstaking research culminated in her report, published by the Centre for Welfare Reform, *What Price Preventable Harm? Social Policies Designed to Disregard Human Need*. This established that Unum (named the second worst insurance provider in the United States in 2008) was brought in as a consultant on welfare reform as early as 1992 and remained a key plank of the welfare reform process by John Major's Conservative government, right through until the implementation of the harsher welfare regime under the coalition government. Her work (which was not funded) revealed the central role of the predatory American insurance industry in revolutionising the British Welfare State from a raft of social insurances designed to give the sick and vulnerable dignity and security into a punitive, twisted Stanford Experiment driven not by medical evidence but by economic ideology, where many staff, socialised into an abusive institutional culture, come to believe they are acting in the public interest by subjecting claimants to cruel behaviourist techniques unworthy of animals.

The evidence also showed that welfare reforms were driving myriad crises across mental health, suicide, food poverty and homelessness. Meanwhile, the UK government sought to suppress information that related to the fact that more than half of claimants in receipt of disability benefit had attempted suicide. Mo also established that the WCA was designed to ensure cost savings by restricting benefit entitlement, and that this was justified by government-commissioned policy research that adopted a new model of determining and monitoring disability and long-term illness. This model disregarded medical opinion and was instead rooted in the assumption that many in receipt of disability benefits were, in fact, 'fit to work'.

In Britain, we have a welfare system that relies primarily on rules and procedures which have no evidential basis. The evidence which does exist points to the fact these welfare reforms are, for many, matters of life and death. Stewart's research proves exhaustively that the British Welfare State is a vast, malevolent enterprise, engaged in the daily manufacture of death, despair and dysfunction, perpetrated from behind an administrative perimeter, fortified by media-generated public ignorance and a political class, spanning all parties, which is either complicit in the cruelty or skimming over the detail.

Mo Stewart will go down in British history as the woman who brought this horrifying truth to light most completely. But there is a broader issue at play here, of which welfare reform is only a part. Their catastrophic impact on many of society's most vulnerable citizens occurs not necessarily because politicians and their supporters have a desire to be cruel, but as a direct result of the low proximity they have to the issues concerned. This, for me, is a far more credible explanation than the common refrain that Tories are simply evil. It is, perhaps, also a more worrying one in that it demonstrates the socially harmful effects of inequality, where different classes exist parallel to one another, with little meaningful interaction between them save for the power one class holds and exerts over the other.

If the voters who support these welfare reforms could see what I have seen, I suspect they might reconsider their view. Sadly, in a society so divided by inequality, many are largely insulated from other social realities. In the absence of any lived experience, they rely on media, word of mouth and assumption in reaching conclusions about issues which fall outside their experiential wheelhouse and, in the process, grant politicians unfettered control over pressing social problems of which they too have a tenuous grasp. It therefore should come as no surprise that across institutions dominated by men from middle- or upper-class backgrounds, the biggest losers by far are working-class, vulnerable women.

*

During lockdown, I interviewed a woman in her mid-thirties, living with mild autism, which was diagnosed at age six, and ADHD, which remained undiagnosed until a few years ago, when it was identified by psychologists. To protect her identity, I will refer to her as May. Due to her learning difficulties, she has been dismissed from every job she has ever had – 16 in total – which include working for an advertising agency and as a cleaner. The longest May has managed to hold down a job is seven weeks. She ended up living in a homeless shelter for a time before being rehoused in a council flat.

May, like countless benefit applicants, has spent hundreds of hours hanging on the line listening to 'that fucking piece of classical music on repeat', which she associates with 'learned helplessness' and a 'constant state of confusion'.

'I don't know if you want to call it direct gaslighting, but the system gaslights you with things like, "we need you to do this to get your money". I was on £53 a week for several years and totally incapable of managing a budget. I was pretty money-hungry, to be honest, and very occasionally literally hungry.'

She describes the 'hoops' through which she would 'dutifully jump', but when she arrived for her appointment – where benefit claimants are subjected to tough questioning about their personal lives, finances and even intimate relationships – she would be given conflicting information. 'I rang the DWP and spent the whole afternoon mainly waiting on the phone listening to that blooming music, being told "Do this", and then when I organised that and rang back, the next guy said, "We would never tell you to do that, that's wrong you must have misunderstood, do this."' May would then jump the next hoop, phone again and be given additional, contradictory information. 'It's the confusion that's the berserk thing,' she explains. 'I felt like I was going cuckoo, and it ends with a total breakdown in trust, and then they wonder why you seem however you seem.'

May's mental health took a turn for the worst as she began doubting herself. Was this just another negative impact of her autism or ADHD, or was the DWP sending her on a wild-goose chase? Often, due to the mental health issues and learning difficulties of many who end up on benefits, the DWP enjoys a comfortable level of plausible deniability where its own incompetence (and malevolence) is concerned. Who's the average person going to believe? A trusted government institution or a poor person who can't even wash themselves? 'It affects people around you,' May acknowledges. 'My mum came to court with me once and she said it was one of the worst days of her life, seeing how they treated me, disbelieving everything I had suffered that she knew full well was true, treating me with suspicion and no humanity. And I feel guilty that she went through that.'

May is a very genuine case of someone whom the Welfare State exists to protect. Her autism and ADHD are documented. Yet, despite being entitled to benefits, her dealings with the DWP caused her to doubt herself, which led to the resurfacing of trauma related to prior abuse at the hands of a former partner. May is a survivor of gender-based sexual violence. 'I was also in an abusive relationship. He didn't hit me or anything like that but I was raped a couple of times over the five years.' The experience of dealing with the DWP, May says, 'made me feel the same: the dismissal, the humiliation, the justifying yourself, your time frustratedly wasted, waiting, with bated breath.' She recalls some of the emotional abuse she suffered as a young woman. 'He used to do fucked up things like hide my shit and then shrug when I went to look for it,' she says. 'I'm not saying the DWP does this on purpose but it puts you in the same state when you need the money to live. They've got your life by the balls. Your mental health depends on their whims.'

May continued complying with the instructions issued by the DWP. Staff insisted that for her money to continue she had to demonstrate that she was looking for work, despite receiving no additional support nor acknowledgement of her employment history as a woman with very specific support needs. 'I was having panic attacks, had labyrinthitis [where you lose your balance from stress], wasn't eating or sleeping regularly at all and I'd had a bit of a mixed time in the shelter, so was sort of recovering from that.'

While managing a decline in her mental and emotional health, May was still being sent to Capita – one of many private companies the government outsources welfare administration to – who would chastise her about how she was dressed, even writing in one report that she must 'come dressed as if going to a job interview' or her money might be stopped. But May was so depressed by then that her personal hygiene had slipped. 'I wasn't even washing,' she says. 'I was crying every day. What the actual fuck? I shouldn't have even been in that office without anyone offering any help. It was just punishment for not living up to standards.'

May did, however, observe that she was treated worse by the private companies than by the DWP staff. Outsourcing is a deliberate attempt to insert a proximity gap, one which creates the dehumanisation required to implement the harsh welfare regime more completely.

After a while, the letters May was receiving just stopped. 'It was a relief,' she explains. 'The appointments were a stressful waste of time, and both parties knew it.' Then, out of the blue, a few months later, someone from Capita contacted her by text, apologising that he hadn't been in touch. He informed her that due to learning he was to be made redundant later in the year, his work ethic had dipped. He said sorry for the lack of contact before announcing he was 'going back to Australia'.

'I never heard from the company again,' she says. 'In the end, I rang the DWP myself to sort it out cos I was worried one day they would stop my money. It's just so sad that, especially initially, I was crying out for help and these people are supposed to help you . . . but there is no genuine actual help, and because they don't know you, they just look at you blankly or deal with you in a business-like fashion – even if you are clearly distressed. It makes you feel lost and crazy, and pretty hopeless.'

In 2019, I encountered Rebecca, another woman in a personal crisis. She had until that point been in a violent relationship with a man. Having seen her posting about it, I contacted her to ask if there was any support I could offer. I knew the guy from the music scene and while I wasn't a close friend of his, it felt appropriate to make the approach.

It was a sadly typical story: she fell head over heels for a seemingly charming guy but over time, he became more controlling and abusive. She suffered regular physical violence as well as emotional abuse – cruel verbal onslaughts designed to obliterate her self-esteem until she genuinely believed she was deserving of such treatment.

Having fled the relationship – in which hard drugs were an ongoing issue for both – she was experiencing what many women do when they attempt to escape an abuser: more abuse. Gradually, she pieced her life back together by getting clean and returning to work. Every now and then, I checked in on her or she would shoot me a message. Some time passed and I assumed everything was OK until one day, I noticed a message she left on a thread asking someone for help to get home from work late at night. I contacted her. It turned out she had recently been raped and was experiencing anxiety travelling to and from her job. The impact of the attack on her mental health eventually led to Rebecca becoming unemployed. Then a run of misfortune threw her into a tailspin. Her landlord passed away suddenly and she became homeless. The stress led to a short relapse which made her even more vulnerable. Rebecca kept going. She found temporary accommodation and applied for Universal Credit.

Disturbingly, for increasing numbers of people, accessing this benefit is where the real problems begin. She became trapped in the hellish six-week limbo between applying and getting money – an arbitrary 'grace period' which is about making the benefit inaccessible – and was forced to access food banks, often spending her nights cold and hungry in hostel accommodation which placed her at further risk of relapse and sexual assault.

For vulnerable women, every possible route comes with a risk attached. Even accessing the services set up to help her was fraught with danger. She shared with me a WhatsApp message from a male addictions worker asking her if she wanted to have sex. Many other men whom she had asked for help demanded sexual favours or nude photographs in return. This deviance has since led to the coining of a new term, 'survival sex', the latest nebulous label masking the behaviour of predatory men who take advantage of women at their most vulnerable.

What disturbed me was how a social welfare system designed to provide a safety net for women like Rebecca, in genuine need, was not only testing her will to live but placing her at risk of more sexual abuse and psychological harm. Until the onset of her adversities she had always worked. A survivor of gender-based violence and an addict in early recovery, she was entitled to support, yet the slightest slip following her assault left her back at square one.

Like me, she fell prey to a regime that struggles to account for

unique personal circumstances. Everyone must fit neatly within a bur-
eaucratic box. If they do not, they can become lost within the system.
But unlike me, Rebecca possessed the additional disadvantage of
being a woman. By then, it was already well understood that welfare
reforms impacted women disproportionately because women often
perform additional roles within families and communities as single
mothers and informal carers but some of the darker impacts, on
those who have experienced the most serious male abuse, are less
understood.

Changes made to the system following the election of the coalition
government in 2010 included cuts to a number of individual benefits,
including child support and carers' allowance, as well as a plan to
reduce the benefits cap – the maximum amount a household can
receive – from £26,000 to £23,000 a year. The Fawcett Society
reported that, in total, 74 per cent of the cuts to welfare came directly
from the pockets of women – and that was before a further £12 billion reduction in spending was announced later in 2015 by Chancellor
George Osborne. More worryingly, the tough new regime of welfare
conditionality put in place around accessing benefits placed victims of
gender-based violence and abuse under increasing strain and at higher
risk because it exacerbated the pre-existing gender inequalities that
make women vulnerable. Just as the system was unable to detect my
circumstances in relation to addiction, mental health and the risk of
becoming homeless, it has proven similarly inadequate where the
safety of victims and survivors of gender-based violence are concerned.

Around the time I encountered Rebecca, another woman informed
me about a recent appointment she had had with the DWP. We will
refer to her as Jane. Called into a compliance meeting – this is when a
claimant is summoned at random to account for some aspect of their
claim, such as an anonymous accusation of fraud or questions about
who lives with them – she had no choice but to take her children
along. At the meeting, she was asked bluntly: 'Who pays for your
children?'

Slightly confused as to the intent of the question, she replied: 'I do,
and my partner contributes.' Her partner, at that time, lived in his
own dwelling, was in full-time employment and was not part of her
household. She wondered whether the DWP suspected her of

committing benefit fraud but as they are not obliged to give any reason for calling compliance meeting, she remained none the wiser.

'No,' said the staff member, 'who pays for your children?'

Assuming she had not been heard, she repeated her answer: 'I do and my partner—'

The staff member cut her off abruptly: 'No – the government pays for your children.'

This is psychological abuse, pure and simple. She was not being asked an honest question; she was being provoked. She was not compelled to the meeting to give information about her circumstances; she was undergoing a form of behaviourist conditioning. After her compliance meeting, she was left to ponder why she had been pulled in. Did someone make an anonymous complaint? A neighbour? A toxic relative? A former abuser?

No mechanism exists within the DWP or elsewhere for a claimant to determine such things. The DWP is an opaque institution, deeply dysfunctional in many regards except one – its proficiency in degrading, terrifying and humiliating those it regards as suspect. But in Jane and Rebecca's cases, we gain insight into a more sinister aspect of welfare reform: the techniques deployed by the DWP against women – coercion, aggression, overbearing surveillance and financial intimidation – mirror those of present or former abusers. Countless women, accessing state support as a result of the adversities they suffer at the hands of abusive men, are then subjected to a similar pattern of abuse by the state (infected by a toxic corporate ideology perpetrated remotely) to which they have turned for help.

In research with victims/survivors of sexual and domestic violence who were claiming out-of-work benefits, Dr Elizabeth Speake, Research Associate at the Department of Sociological Studies at Sheffield University, in a 2020 doctoral thesis titled '"Like having a perpetrator on your back": Violence in the Welfare System' found many similarities between women's experiences of abusive relationships, and their interactions with the DWP. Part of the research focused on the process of applying for and claiming disability benefits, a process which was experienced as humiliating, violating, and (re)traumatising by the women interviewed. Following a request for a summary of her research, Speake wrote:

'The anticipation and aftermath of assessments, in particular, often led to significant relapses in mental health issues such as PTSD, depression, and suicidal thoughts. Women identified several specific parallels between these situations and prior experiences of abusive and controlling relationships, including disbelief, "gaslighting", invasions of privacy, hostility, and sometimes outright aggression. Furthermore, as each of these interactions were underpinned by the threat of further loss of resources and therefore, choices, the assessment process produced a profound sense of powerlessness. In a situation like this, where there is a significant power imbalance, and one side is using their power to elicit deeply personal and traumatic information from the other, with their demands backed up by threats to withhold the most basic resources, a clear comparison to coercive control can be drawn.'

Most importantly, several women in the research explicitly characterised their relationship with the benefits system as abusive, with one stating, 'It's like having a perpetrator on your back [. . .] yeah, control over your money, emotional, mental, psychological abuse'.

In recent years, a growing body of evidence and testimony points not simply to victims of domestic abuse struggling to access benefits to which they are entitled, but also being subjected to further torment and placed at more risk by their mere interaction with the Welfare State.

In June 2019, a welfare worker named Elaine wrote a guest blog on the Trussell Trust's website raising the alarm that government departments – while outwardly keen to support victims of domestic abuse – were in fact exacerbating it, placing victims at further risk. Elaine wrote:

'Unfortunately, what we are witnessing are life-threatening situations involving domestic abuse further exacerbated by issues with the government's new benefits system. The lengthy period of time between making a Universal Credit claim and receiving the benefit impacts not only on women's ability to support themselves and their children, but also on the legal process they are often going through when escaping domestic abuse.'

Elaine also noted that where victims have sought refuge after attempting to leave abusive partners, they must also contend with a legal struggle to retain custody of their children. One that requires legal aid – a means-tested benefit which has also been cut – which can only be granted when proof of benefit entitlement is provided. The six-week lag between applying for Universal Credit and receiving it means this proof is often not available, leaving victims and their children vulnerable not simply to the acute financial hardship of weeks without income but also to additional abuse in the family courts where they must often face their abusers without legal representation.

'This creates a nightmare situation for survivors,' Elaine argued, 'who, having taken the terrifying step of escaping their abuser, are then expected to fight for their children's safety alone in the family courts.'

May, Rebecca and Jane are part of a worrying trend. As are John, Glen and the boys from Pilton. The Welfare State no longer exists to protect working people from the pitfalls of the free market – it has become infected by a toxic corporatist ideology that is behind its worst excesses. An ideology where citizens become assets and liabilities in public services which are run like cash-and-carries – except, where the poor are concerned, the customer is never right.

Austerity was a sham. The financial crisis that preceded it was created by the same political class and private enterprises which then oversaw the savage cuts to public services that have, according to a report by the IPPR think-tank, been linked to more than 130,000 preventable deaths since 2010. The explanation that this occurred because politicians are evil does not satisfy me – that would imply competence. The rather banal truth is this: the social distance between those in government and those on the sharp end of inequality cannot be bridged by a few photocalls in poorer communities. It has become so vast that even when some politicians have acted with the intention to improve people's lives, they have instead immiserated them – or pushed them over the cliff's edge.

But it's the hypocrisy of it all that should really anger us. The blatant double standards of a political class, which has for the longest time enjoyed a parallel benefits system which not only distributes public money to them far more generously, in the form of subsidies, expenses and other entitlements, but also one where the rules in place around it are nowhere near as punitive.

9

I Don't Like to Be Beside the Seaside
Immigration concern and the political mismanagement that drives it

It's July 2019, on the south-east coast of England, in the town of Margate. The basic ingredients for a booming seaside economy are present all around me: miles of well-pathed coastline, fresh, clean air and a shimmering sea stretching out to the horizon. Yet, somehow this town appears to be in the grip of wicked depression, chained in restraints, tragically unable to rise to the seasonal occasion which has historically shaped and defined it – summer.

I arrive at my hotel to find the reception area empty. I ring the bell and a friendly woman approaches, apologising for not noticing me. She seems out of practice. We do the obligatory 'Oh you are from Scotland my friend lives in Edinburgh' thing that all Glaswegians must endure when travelling any further south than Gretna Green before I am handed a key and directed towards the lift. Ascending three floors, I step into a narrow hallway. The sound of daytime television can be faintly heard nearby. It's been a while since this place saw a lick of paint or a square-foot of fresh carpet by the look of things, and quite some time since I've had to bear down in such basic accommodation. The mild agitation caused by trying to squeeze myself through a rather slender doorway and round a tight corner to my room serves as a welcome reminder of how pampered I am these days. Not long ago, I'd have happily slept on a bed of beer cans and fag-ends – now I'm a pompous hotel snob!

By morning, I'm raring to go as I arrive at a local community centre which is home to some of the youth services that offer activities aimed at drawing young people away from anti-social behaviour and crime.

In recent years, services have been cut by the local authority, leaving youth workers exasperated as a constant drip-feed of news stories about the 'knife crime' crisis dials up the anxiety among many young people, often leading to them carrying weapons.

I detect a hint of scepticism from the Quarterdeck Youth Centre's manager, Albert Edwards. After all, I'm an outsider with an agenda, and he hasn't quite had time to suss me out yet. Still, he's kind enough to humour me as I pose my first few nervous questions. It isn't long before the conversation turns to racism. 'I think it's bred; it's bred for generations. In smaller communities, I can go back as far as 1960, you'd hear the white father next door say to his son, go to the "paki" shop, go to the "chinky" shop, and it starts from there. If you've got a mother or father putting down another race and you've got the kids hearing that, how they gonna come out? How is it ever gonna change?' Albert tells me that changing racism is about more than just calling out individual instances of racism in the community, it is also about confronting those in power at local and national level, before alluding to the racism at 'KCC' – Kent County Council.

Political representation for ethnic minorities is a real problem outside the large urban centres in the UK. In Kent, Conservatives dominate the political landscape at local and regional level. Naturally, they are inclined to look after the concerns of the people who vote for them – most being white – which means issues impacting the numerous ethnic minority groups are hardly a priority.

Joining us outside is my tour guide for the day Ayaan Bulale, a community artist who is of Somali descent. 'When you are an immigrant, you tend to vote Labour, so when there's a government like we have currently then the disenfranchisement and disengagement becomes even greater. Some of my friends didn't vote in this election. They can't cope. But what are you showing your child if you are not voting? I lost my vote because I'm an EU citizen, so I'm not even allowed to vote for the EU elections. They took away my fucking vote. I called the newspapers. I kicked off everywhere. Spoke to my local MP.

'My parents were economic immigrants who came to Sweden and Denmark in the 1970s. Even though it had one of the best welfare states in the world, my parents were like "We're not getting anything for free. No, you have to integrate. Go outside, don't just hang around with other Somalis – and speak Swedish." When we come home, we're Muslim so were speaking Somali. But when we step outside, we got to do whatever we wanted to do. My parents let me do what was, for a girl in their culture, inherently wrong – wear shorts, play with boys, not go to mosque. So, I was taught "integration is amazing, assimilate". Your parents' attitude when they arrive, and what they instil in you, in the same way we talked about racism, they teach you that and that's the path.'

Ayaan's case is an interesting one. She talks up the pros of adapting to a nation's culture when you arrive as a migrant. She's firmly on the left, politically, but this emphasis on personal responsibility, even with respect to a protected group like an ethnic minority, shows an independent mindedness which I'm certain rubs her comrades up the wrong way from time to time. Whether her parents promoted assimilation because they truly believed in it or it was a strategy they adopted to avoid racism is unclear. Either way, she represents a view on immigration which is not often accounted for in the debate: a woman of colour, from a conservative background, since turned progressive, who believes immigration is a force for good but understands the importance of community cohesion. Only by visiting an area where immigration is a live issue and speaking to people on the ground does this ever occur to you. You may wrongly presume that all ethnic minorities would be in favour of the open-borders policy advocated by many liberals and left-wingers, but you'd be wrong. And just as much of the debate on the left of the spectrum around immigration revolves around the binary of the migrant's plight versus the hostile white ethnic population, immigration concerns themselves are often similarly misunderstood.

Despite the perception among some that Britain is inherently anti-immigration, immigration concern among white British people has declined overall since 1964, when attitude surveys on the issue were first introduced. Throughout the 1960s and 1970s, as many as 80 per cent of people responded 'Yes' when asked if they believed too many people were being 'let in' to Britain. From this peak in immigration concern, attitudes softened somewhat until the turn of the millennium when Freedom of Movement led to increased numbers arriving from newly admitted EU countries. In the year or so before the EU referendum, between June 2015 and June 2016, immigration was consistently regarded as the most urgent issue facing Britain, peaking at 56 per cent in September 2015. Political parties rose and fell in the polls based almost entirely on either their immigration policies or the perceived strength or weakness of their statements on the issue. But since the EU referendum in June 2016, the Migration Observatory's analysis of immigration attitudes notes that 'immigration has been mentioned by far fewer people, falling from 48 per cent in June 2016 to 13 per cent in November 2019', when concern about immigration was supplanted by concerns around Brexit which in November 2019, 62 per cent mentioned as a primary concern. Given immigration numbers had not shifted tremendously by 2019, the argument that immigration concerns correlate with migrant numbers makes little sense. Indeed, fears and anxieties around immigration appear to correlate more closely with the level of polarisation in discourse, and the substance and tone of news and media coverage, both of which are always preceded by growing levels of inequality.

I say my farewells at the youth centre and am accompanied by Ayaan for a brief walk around the community. My time here is short and she is able to negotiate my entry quickly. We cut immediately off the main street and up a rather manky lane. 'We have to go the long way,' she tells me. She is going to check in on a local family of Eastern European descent, living on the ground floor of an accommodation they are renting from a local landlord. 'We come the long way because the landlord built this wall,' she tells me, pointing to the large barrier which, she explains, was designed to prevent the migrants living here being seen from the main street. 'It adds value to the property,' she says, the disgust faint in her otherwise dazzling eyes, 'so they have to walk all the way around the block to get in and out of their home.'

Some children run out of the house to say hello. Despite my having no idea what is being said, it is clear from the body language of the children, and the fact no parent has emerged to investigate, that she is a friend of the household. We say our goodbyes and trudge on, the sun beginning to bake my pathetically pale skin – and the unemptied bins which dominate large areas at the back and front of many buildings.

As we walk, she describes the local housing market. 'Move to London, your parents help you buy your first property. You sell that and now you've made your money. My friend bought one, a proper Margate property, two bedrooms. She paid £65,000 in July 2017 then sold it to her hipster friends. All she did, I swear to you, was hoover it, paint the windowsills and put up this one set of blinds, then sat on it for a year. It sold for £135,000 one year later.'

Despite my awareness that the stupidity of many hipsters is boundless, I am stunned. For a property purchased at such a low price, modified only slightly, to then sell for almost double the cost makes very little economic sense – unless, of course, you're the one turning a profit. 'When you look at the quality of the architecture,' she says, gesturing at the prestigious facades above head, 'these were all guesthouses which were then split into multiple-occupancy homes.'

Globalisation is often discussed in the context of the modern world but for many coastal towns, where employment and social mobility were dependent on tourism, the downturn began a long time ago. Exacerbating the sharp economic decline in recent years is the overspill of Londoners seeking to escape the high cost of living in the capital. In places like Thanet, the local government area in which Margate is situated, Londoners get far more bang for their buck, but the influx has fundamentally altered the town's economic topography, creating a landscape where areas of deprivation and affluence sit side by side yet appear worlds apart. The level of gentrification that has taken place is, like most areas, uneven: some parachute in with a sense of entitlement which places them at odds with the local culture while others are more respectful, attempting to embrace the local community.

We take a left off the beaten path, entering a street which, at first, feels strangely partitioned from the surrounding area. The energy is different, like some of the oxygen has been sucked out of the air. On both sides of the road, there are tall, white tenement buildings with long, grand staircases stretching up to the doorways. It's now midday and some residents have pitched up on the pavement. From a distance, you'd be forgiven for thinking they are loitering and, given they are all young men, keeping your head down wouldn't be unwise. But as I approach the group, gathered on the pavement at the foot of the stairs, I discover one of the men is a mechanic. I stretch out a hand to say hello but he politely declines due to being covered in motor oil. The other men gesture hello, my passage through the community assured on account of my guide, who seems known and loved by everyone we encounter.

'You never know what you can expect on each floor,' she says, as we enter the building, which is carpeted. A pungent unpleasant stench hangs in the air that catches the back of my throat. 'Some people make an effort, some people don't,' she remarks, as I observe the cramped, unkempt conditions. The immediate aspect of the building that jumps out is how unsafe it feels. It's much like entering a derelict structure, except instead of a deathly silence you can feel the low hum of human life. 'Where's the safe fire exits? How do you get in and out? There is no light,' she tells me.

I comment on the mess of the street outside, which is undeniable, but she explains that while unsightly bins have created an impression that migrants are unclean, the issue often stems from there being no designated bin areas – refuse is placed in front of the dwellings for collection. Collections which, as a result of austerity, have been cut. 'You have eight flats here, eight flats there, two on each floor. Twenty-six bins. Three big bins. So then people go, "They're standing there, standing there next to the rubbish, immigrants, rubbish, they cause the problem, look how they live."'

With the near-deafening cackle of seagulls overhead, it becomes clear why much of the rubbish ends up scattered around the street – at night, the bins are raided by wildlife for food scraps. This creates the conditions for a broken-window effect – broken-window theory states that visible signs of disorder in an environment create further disorder leading to decline in a community – the likes of which many in Britain will never experience but nonetheless will develop strong opinions about. Opinions which many canny political operators have appropriated for their own ends.

This, in part, explains why districts like Thanet became lightning rods for the Brexit debate. They present outwardly as run-down incubators of poverty, unemployment, crime and social immobility. Eastern European immigrants, who have travelled here legally under Freedom of Movement laws, are then blamed by some locals, not simply for the rubbish, but for being symbolic of the socioeconomic malaise which has gripped their community. Migrants are simultaneously blamed for taking jobs and for being on the dole. For gentrifying the landscape and for sitting around all day and doing nothing.

While to many, immigration (and the EU) appears a central factor in the declining socioeconomic fortunes of their areas, the problems of ageing populations, low educational attainment and unemployment, as well as poor economic fundamentals – exacerbated by austerity – are long-standing, preceding Britain's membership of the European Union and attributable to governance at local and national levels in the United Kingdom.

In areas like Margate, it is not just the plight of migrants which is misunderstood; immigration concerns and anxieties about immigration are similarly obscured by the polarised discourse they generate. Following the terms of the debate as it exists in the social media space or in the mainstream press, you may come to an understanding of immigration in the UK which is no understanding at all. The first counterintuitive fact you will contend with, in the real world, is that there exists a broad diversity of views on immigration, not least among migrants and people of colour.

The 2013 British Social Attitudes survey posed questions about immigration. Analysis of the findings revealed that the majority of white and BME (Black, minority ethnic) respondents favoured 'at least some reduction in migration levels', with only slightly less agreement among those born overseas. This shows that there was broad public support for the view that the level of immigration in 2013 was too high. This was expressed in the general election of 2015, when support for the Conservative Party among ethnic minority groups increased on the levels seen in 2010.

After the 2015 general election, British Future/Survation polling of ethnic minorities found that 33 per cent voted for the Conservative Party – considerably more than the 16 per cent who voted for the Tories in 2010. And while this proportion remains well below the 52 per cent of the ethnic minority vote which Labour gained, ethnic minority support for the Conservative Party did increase in the years that divisions over immigration became more pronounced. Ultimately, as the Tories positioned themselves as the party tough on immigration and antagonistic towards the European Union, its support among people from ethnic minority backgrounds increased.

A report published in 2015 by Conservative think-tank Bright Blue, titled 'Understanding How Ethnic Minorities Think About Immigration', found that 40 per cent of those polled from ethnic minority backgrounds desired 'a system that is well managed and efficient at keeping out illegal immigrants'. Prioritising immigrants who will contribute positively to Britain's economy and social fabric was also regarded as important, with 25 per cent saying those should be the main characteristics of an ideal immigration system. In the nationally representative sample, ensuring a well-managed system (35 per cent) and only admitting immigrants who will contribute (24 per cent) also came through as the most popular characteristics of an ideal system. As noted by the report, 'Competence and contribution are characteristics which are prioritised across the population.'

This is, of course, based simply on attitudes to immigration which are not always shaped by facts. The regimenting issue around Brexit was EU immigration – that's what millions were led to believe was the problem. The 'Hostile Environment' reforms – immigration legislation designed to target people in the UK illegally – were so harsh in part because, as a result of Freedom of Movement, the UK government couldn't regulate how many people came from Europe. So, in order to meet its own ill-thought-through manifesto commitments on immigration – which included a pledge to cut net migration to 100,000 – non-EU immigrants had to be pursued more aggressively. By pinning its credibility to an issue it had little control over and by pandering to falsehoods about EU migration for short-term political gain, the Conservative government cleaved a great distance between the debate about immigration and the economic reality – Britain cannot function properly without sufficient immigration levels and immigrants make a substantial economic contribution to the public purse.

In preparation for Brexit, the government requested a report from its Migration Advisory Committee (MAC) on the economic and social impacts of EU migrants in the UK. The MAC commissioned Oxford Economics to analyse the fiscal implications of immigration using the most up-to-date data and sophisticated modelling techniques. The resulting study, *The Fiscal Impact of Immigration on the UK*, found that 'the average UK-based migrant from Europe contributed approximately £2,300 more to UK public finances in 2016/17 than the average UK adult' and that in comparison, 'each UK-born adult contributed £70 less than the average'. Notably, the average European migrant arriving in the UK in 2016 was projected to contribute £78,000 more than they took out in public services and benefits over their time spent in the UK, and even the average non-European migrant was projected to make a positive net contribution of £28,000 while living here – the average UK citizen's net lifetime contribution in this scenario, as noted by the report, was zero.

Essentially, the EU migrants who arrived in 2016 would go on to make a total net positive contribution of £26.9 billion to the UK's public finances for as long as they remained here – equivalent to putting approximately 5p on income tax rates (across all marginal rate bands) in that year.

Still, this net contribution does not undermine the broader point about the complexity of immigration concern, particularly among people from ethnic minority backgrounds, which must be understood in order to truly grasp the cultural dynamics of communities where concerns were more pronounced. 'Net' contributions, like everything else, are unevenly distributed, meaning people at people at the lower end of the income scale do not experience them in the same way those near the top do. Distance from the spoils of immigration, and simultaneous overexposure to its complexities, means areas with high levels of immigration and low levels of public investment are bound to become breeding grounds for anxieties and concerns. Then, of course, the vast socioeconomic and cultural gap between working-class people in those areas and the cosmopolitan middle classes in the big cities observing them from a distance means immigration concern is framed as an issue of white working-class people. And so, an additional misalignment between facts and discourse emerges.

The UK's leading race equality think-tank The Runnymede Trust, which (as well as helpfully collating much of the research from which I have generously drawn in this chapter) found, in 2015, that while different experiences drove some of the immigration concerns of white ethnic and ethnic minorities, 'as with the wider population,

Black and minority ethnic people see some positives and some negatives for immigration to Britain'. The report noted that 'where BME people are concerned about levels of immigration, this is more likely to focus on the fairness of benefits, or the pressure on social welfare policies' as opposed to cultural or security concerns. These anxieties and concerns are then compounded by the understanding that 'fairness' arguments about access to school places, housing and maternity services may imply that they and their British-born children have fewer rights to access public services, or that they increasingly have to 'prove' they are actually British and thus entitled to access public services and benefits.

The report also highlighted that BME people with concerns about immigration also feel a parallel concern that when the topic becomes the subject of fierce debate, they, as people of colour, may get caught in the crossfire – despite being either British or legally settled here. The views held by ethnic minorities in key areas, such as on the pace of change and contribution, do not differ greatly from those held by the wider population, despite the vast differences of experiences in Britain between white ethnic and ethnic minority groups. The prevailing notion among many left-wingers and liberals that immigration concern is driven by hostility to immigrants because they are foreign is laid bare. The desired attributes of a well-integrated migrant tend to be the same for ethnic minorities and the wider white ethnic population. Though many on the left who argued that the debate was a red herring, arising from bad-faith arguments and heavily biased right-wing media coverage, are entirely correct.

Nonetheless, the failure of those leaning more left on these issues to interpret the complexity of concerns about the pace of change was a boon for the far-right. It is no surprise that in 2015 – one year before the Brexit vote – Thanet District Council fell to UKIP; this was the same year as many as half of the children in some areas of Thanet were found to be living in poverty. In the absence of rational discourse around class inequalities and, indeed, EU immigration, the vacuum was filled by Nigel Farage, who targeted post-industrial areas like Margate with his political campaigns, retrofitting historical policy decisions taken by British politicians and civil servants about how communities should be structured to a caricature of the distant

European bureaucrat. This message stuck because it was simple – despite Kent being among those UK counties to receive a net benefit from the EU. Margate benefited from a £900,000 investment in coastlines development and 127 nurses who work in the NHS in the town came from EU countries. But the framing of Margate's decline as a direct consequence of overbearing EU institutions felt intuitively correct for many, despite having little basis in evidence. In truth, the evidence points to the disastrous planning and low levels of investment in such areas for decades, much of which was overseen by Conservative councils. Farage himself even parachuted in at the 2015 general election and was projected by many to win the seat, but as a result of the Conservative Party adopting his rhetoric, it instead went to a Tory.

Thanet District Council has long been dogged by accusations of mismanagement and incompetence. A damning report into the local authority in 2013, published by the authority's standards committee, painted a picture of an institution largely distrusted by the public. It even noted how, on some occasions, the public was treated with 'outright hostility'. The report states: 'There is a local suspicion of secrecy, corruption and distance between the council as it is perceived in the offices in Cecil Square, from the reality of people's lives and the needs of the district. The council has the appearance of a dysfunctional organisation whose behaviour and internal squabbles adversely affect the delivery of services, capital projects, etc., to the residents of the local district.' It concluded that behaviour fell short of the council's own standards and was in urgent need of 'rehabilitation'. Sadly, the poorly run local authority represents just one dimension of the area's chronic social and economic problems.

Back in Margate, the curdled social media discourse around immigration feels almost trivial. It is perfectly obvious why some are content that the focus remains squarely on immigrants and not governance as we traverse the many chaotic hallways of the dangerous building, before arriving at the home of a young Eastern European family, who invite us in for a cup of tea. It comes as a welcome sensory reprieve from the oppressive mess outside. Three children are seated at a table, quietly eating lunch. The house is immaculate, making efficient use of the limited space. Given the midday heat, the windows are thrown open – it finally feels like summer.

'Water?' asks the lady of the household (whose identity I will protect). I decline but only because I'm so caffeinated that any more fluids would require multiple toilet trips and I don't want to be that guy. Her partner is dressed in shorts and t-shirt, sucking on the end of a vape pen – a greenlight for me to use mine. We take seats in the living area and I explain why I'm here, that I'm keen to understand what life entails beyond Scotland and that, given Kent's reputation as a wealthy place, how does that line up with people's experiences of poverty. The woman is Czech, the man Roma, and their relationship is a sticking point for many back home – their romantic union breaches firm lines of custom.

'A Czech person is not supposed to be with a Roma person,' says Ayaan. Roma people count themselves among Europe's most persecuted. The Romani community makes up 2 per cent of the Czech Republic's 10.6 million inhabitants, similar to the wider European average. There are 6 million Romani living across the European Union and studies consistently show that there is not one country where they are not discriminated against on multiple fronts. Wherever they reside, they face prejudices, chief among them, the battle for basic housing. 'The fact they are together is a miracle,' says Ayaan.

I ask the woman how she feels about being regarded by some of the other locals as unclean. 'I don't have any reaction because I know myself, so I ignore them.' Are people outwardly racist towards her or her family? 'No,' she replies firmly, her partner concurs. Ayaan expresses surprise at their response, but having spotted the weight bench in the back and the fact her partner is built like a brick shithouse, I think there may be a simpler explanation.

'You don't hear racism around here? Cos I hear it all the time, especially when I'm in taxis.' She then breaks into a wonderful impersonation of a racist cabbie, which, while veering a little on the side of stereotype, delights everyone.

'When we came here years ago, the people [she gestures to indicate they were mistreated] but now it's different. We have good, nice house, but you have to clean every single day.' I feel her pain. With only two children at home, I am already aware the housework is overwhelming. She has quite a few more than that, which means her work is never done. Has life grown easier as they have assimilated?

'Sometimes it's easier, sometimes it's hard. You up and down. When there is working it's good.' Her partner chimes in, nodding in agreement. They came to the UK in 2005 to find employment and escape the persecution they faced. They picked apples for three months before falling pregnant. She stopped working and he became the breadwinner, but work has been hard to find lately.

In conversation, I become aware of my own preconceptions and how, even from a place of compassion, I am forming judgements. Why does it surprise me that their house is spotless? That their children are all immaculately turned out and annoyingly well behaved? Why do I keep trying to bring the conversation back to the bins outside? Has my lack of proximity to the issues families like theirs face made me prey to assumptions, bias and even prejudice? I thought I was one of the good guys, yet even I, as a result of my distance from this issue, can fall victim to the same lazy tropes I'd identify as racist or xenophobic in others.

The family rents from a notorious letting agency. 'When I move here five years ago, they charge £50 for my post,' she tells me, referring to the practice of hitting tenants with excessive and arbitrary charges. In this case, to obtain their own belongings from a previous address. 'They said, "It's not your house now" and I have to give them money but I had no money. I lost everything.' The estate agents changed the locks while the family was mid-move, preventing them from accessing the property.

'They wouldn't do that to you and me,' says Ayaan, 'but if you don't speak English they can turn up on the day you are moving out. It's a racket.' The only way out of the nightmare of private letting is to get a council house, or buy a home, but with a chronic social housing shortage, safe, secure public housing is scarce.

The housing market is also largely inaccessible either through high costs or tough loan conditions. In the absence of secure work, a guarantor who earns at least £30,000 is required, as well as the £25,000 deposit. When the family moved into their last accommodation, they paid £1,200 in rent up front, a £1,200 deposit and were charged an additional £450 in admin fees – an illegal charge which many letting agents openly levy because vulnerable sections of the population don't know their housing rights. The woman describes a recent rent rise applied with less than a month's notice – also illegal. When these practices are applied at scale, across large sections of the migrant population, they generate large windfalls which create further incentive to bend or break the law. The rental market creates a moral hazard where dodgy letting agencies are prepared to act in blatant breach of the law because the promise of profit outweighs the risk of punishment – deterrence becomes inverted.

So where do the family go for support? Who are their political representatives? Sadly, she does not seem to understand the question. The concept of political representation, as well as the notion that anyone may advocate on her behalf, is as foreign to her as she is to some of her neighbours. 'Four years you have to wait for a council house. I've been here for 15 years,' she tells me, laying bare a central myth around which anti-immigration sentiment is cultivated: a chronic shortage of affordable social housing forces people into a cowboy private let market, creating a well of resentment within communities where some wrongly attribute declines in housing conditions to the presence of migrants – not poor local governance and corruption.

The immigration issue is riddled with problems of proximity – many of them we have already established in relation to other areas. It was around Brexit that proximity gaps effortlessly converged, in a perfect populist storm. There was the educational attainment gap between Leave and Remain voters. The jobs and professions many in these respective groups occupied, the subsequent social status and economic security (or lack of) derived from employment and the contrasting levels of labour market disruption each were exposed to. At play too were the differing levels of trust in authority and public institutions. Much of the working-class Leave vote was mobilised not simply by hitching the campaign to the lightning rod of immigration, but also by tapping into a well of prior resentment and scepticism of governance, arising from lifetimes of negative experiences of authority and public services in poorer communities.

In contrast, many middle-class Remainers, as a result of the economic confidence associated with their social position, could barely comprehend the ulterior cultural forces driving the Leave campaign's groundswell of support and failed to recognise that Brexit was not really about immigration, or lies about money for the health service, or flags – it was about retribution for decades of being ignored and spoken over. That prominent pro-EU voices could not see this and instead chose to portray Leave voters as either gullible or xenophobic, spoke to how remote they had become from large sections of the country. They, in effect, appeared to exhibit a deeper fondness and sense of kinship with continental European bureaucrats in the aftermath of the referendum than they did with other Britons – a mark of the gulf between Britain's social classes.

The most significant factor by far, however, was surely that the Brexit debate was ever centred on immigration in the first place. That was perhaps the ultimate red herring. Not least because the biggest predictor for immigration tensions, as evidenced in Margate, is not the number of migrants in a given community, nor events occurring internationally. Immigration concern is driven, always, by shoddy local and national governance.

10

A Deferent Class
Knowing your place

The lopsided distribution of wealth and the opportunities and margins for error this affords is the essence of class inequality. The dissimilarity in the economic interests of those on either side of the social ravine forms the basis of class conflict and struggle. But baked into these more measurable social facts are subtleties and sensitivities so elusive that they often escape our notice. It is already well established that social class plays a role in determining how long we live, how well we are educated, what papers we will read, what we'll do for work and how much we'll be paid for it. But ulterior to those areas of divergence lie cultural differences which, while certainly palpable, are trickier to measure and even tougher to articulate.

Now that we have covered the main structural barriers (at least as I see them), we shall, over the next few chapters, examine subtler cultural gulfs between social classes, developing some of the themes we established at the end of the previous chapter, and attempt to parse out their implications.

A key area of divergence and one which is central to understanding the cultural impact of structural class inequality is trust and reverence for authority. Gulfs between social classes in this area were exposed during the pandemic, in relation to public health advice, mask-wearing and, perhaps most contentiously, the vaccine rollout.

Consider the variations in vaccine uptake and the fact that people who were most vulnerable to Covid were also often the most hesitant to get jabbed. Rates of vaccine hesitancy were lowest for working-age adults with higher incomes and highest for those in the most deprived areas. Homeowners were less hesitant than renters. People educated below degree level were more hesitant than those who went to university. And Black or Black British adults had the highest odds of reporting vaccine hesitancy when compared with white adults. Isn't there a theme here?

While fake news and conspiracy theories ran amok during the crisis, by far the biggest reason people were hesitant about the vaccine was their concern about side effects – a valid concern. But beneath that concern lay arguably a deeper scepticism of health advice. And beneath that, a distrust of authority, generally. Despite the reassurances of medical experts, and the repeated encouragement from political leaders across the spectrum to get jabbed, millions have yet to take the government up on the offer of vaccination – including many who work within the health service. Are those people stupid? Perhaps some of them are. But is a more realistic explanation not that prior negative experiences of public services and authority informed their individual and community attitudes to it? Is vaccine hesitancy a sign of idiocy or simply the cultural karma which has accumulated over time in a country where some people are treated better than others, creating a well of distrust, making likelier the formation of sceptical attitudes towards public health advice?

Nations such as the Scandinavian countries, New Zealand and Australia, while still suffering greatly during the crisis, benefited from their comparatively higher levels of public trust, granting governments greater flexibility where health advice and public restrictions were concerned. Unsurprising, perhaps, that those countries are better governed than Britain's home nations. Interestingly, in the UK, the same groups among which vaccine hesitancy was higher were also those likelier to have had negative experiences of education, healthcare and housing; they were also likelier to have found themselves on the receiving end of authority, namely police involvement. For many, the vaccine rollout represented perhaps the first time in their lives that a public authority had actively reached out to them with more than a threat or ultimatum. The pandemic was, for many, the catalyst for the state taking a sudden, suspicious interest, after a lifetime of appearing either disinterested or antagonistic.

This dramatic shift in the government's general tone and posture towards those on the lower end of the scale was perhaps best exemplified by how rapidly rough sleepers were housed – not for their own benefit but to break the chains of transmission that may have led to the NHS being overwhelmed.

It's easy to sneer at those who refused vaccines or who expressed concerns. And the anger towards them is doubly understandable when it comes from those directly impacted by coronavirus. But we must, even in our legitimate anger, consider the depth of conviction required to refuse it and the political and economic roots of such hard-line, intransigent attitudes. These were often people who understood the danger. Who did not doubt the virus was real. Given many of their prior experiences of authority, it is not hard to see why some would question the motives of public officials, even if it meant social ostracisation, cultural ridicule and even serious illness or death.

When distinct communities emerge in different material conditions, shaped by distinct economic contexts, their beliefs, attitudes and priorities are bound to diverge, just like their experiences do. The further down the social scale you go, the harsher the experiences. At the extreme end, there is not just distrust of authority but absolute irreverence and even hatred for it, in all of its forms. But this vast culture gap between social classes is most pronounced with respect to experiences and thus attitudes towards law enforcement.

Nobody likes a grass, or so the saying goes. This mantra is pervasive in highly policed communities and is often assumed a mark of idiocy, mischief or malevolence from a distance. Where I come from – and particularly in young, male communities – the only thing worse than a 'grass', 'rat' or a 'snitch' is what you will happen to you if you are ever accused of being one. In prisons, those thought or found to have cooperated with the police are regarded with the same disgust as paedophiles and often live in 'protection', like sex offenders.

In this moral universe, there are few crimes that justify snitching. This is how lowly the criminal justice system is regarded. When views of the police have been shaped not by respect or reverence for authority, but by being constantly subjected to it, to confer with law enforcement in any manner is to knowingly ally yourself with the enemy. People adhere to this belief with different degrees of commitment but even moderates will become extreme in the face of this social pressure. From this social vantage point, which is certainly more common in communities where poverty, and therefore crime and the presence of law enforcement are greater, not grassing becomes an ethically legitimate, philosophically coherent and socially responsible position, and signalling it to others further cements you as a person of good standing.

This is not to say that 'don't grass' is always invoked virtuously. It is often parroted most aggressively by 'characters' in the community least worthy of trust or respect. This philosophy forms precarious ethical terrain from which they may fashion for themselves a moral high ground from which to preach their own brand of overbearing authority. And many others who live under the tyranny learn to repeat such nonsense reflexively, to keep people like that off their backs.

What must be understood is that while this may appear a strange moral world to inhabit, it is a moral world nonetheless, and must be regarded as such by anyone hoping to meaningfully engage with people who hold these beliefs deeply. These values arise in communities where certain individuals and groups disproportionately find themselves on the receiving end of authority – irrespective of their guilt or innocence.

Criminal justice systems have punished the poor disproportionately since their earliest iterations in the ancient world. The punishment for committing a crime in Rome was not the same for everyone but was tied to social status. Wealthy patricians received more lenient punishment than slaves for committing the same crime. Today in Britain, men and women of higher social status are under-represented in the criminal justice and prison systems. Wherever they do surface, they are often spared the harsher punishments handed to lower-class people because it is believed that they would not 'benefit' from custodial sentences. And some fully paid-up members of Britain's upper class, whose position in British society is entirely hereditary, are not even required to submit to legal investigations concerning the most serious, lurid and depraved allegations.

If you're still having trouble with the notion that class is real, you need only set foot in a British prison for a few moments and behold for yourself the predominance of one social class to see who British criminal justice systems are primarily designed for. At the heart of the 'don't grass' mantra is a basic and legitimate concern about class. An unconscious one, perhaps, but a concern, nonetheless. The 'don't grass' philosophy is rooted firmly in a recognition that as a lower-class person, in a country where the police cells, courts and prisons are full of people who look, dress and sound like you do, your interests may be better served by not cooperating with law enforcement at all – whatever the consequences. Meanwhile, at the opposite end of the social scale, police exist mainly to protect and reassure people.

It's a bitter but bright enough morning in New Cumnock, East Ayrshire, in the south-west of Scotland, and I'm preparing to meet one of its most famous sons, Sir Tom Hunter. The businessman and philanthropist, who made his fortune selling working-class youths the tracksuits, baseball caps and trainers that defined their generation, has kindly agreed to an interview for my BBC Scotland film series, *Darren McGarvey's Class Wars*. Despite my considerable preparation, I can't help but feel slightly nervous. My producer opens the Covid-secure car to inform me that I am being moved to the 'other' car. By 'other', she means the colder, smaller and altogether less comfortable car. The news is cause for mild distress as I have become accustomed to the superior vehicle. You think you're a big deal until a billionaire walks in the room – or in this case, a car park.

Before Tom's arrival, I nip into the Tamar Manoukian New Cumnock Pool for a tactical toilet stop. Refurbished by the Dumfries House Trust in 2017 and counting Hunter as one its sponsors, the New Cumnock outdoor swimming pool, opposite the New Cumnock Town Hall, is a bright, pastel-coloured oasis in an otherwise remote, aesthetically underwhelming and economically underdeveloped former mining village.

Cumnock is one of constellation of towns in the south-west of Scotland which expanded aggressively throughout the coal-mining era in the late eighteenth century, before falling on hard times with the closure of the pits – around which their economies, cultures and social bonds were forged. Despite the usual attempts to place some much needed distance between the town's former reputation as a run-down area, high in unemployment, and its budding twenty-first-century makeover as a beacon of regeneration and historic significance, you can still feel here, in the Cumnock air, that resignation and fatalism so palpable in these ghost towns. Local authorities are keen to present a picture of economic and social progress, in a bid to attract investment from housing developers, supermarkets and multinational coffee and junk-food chains, but the child poverty statistics tell you everything you need to know about the efficacy of this orthodoxy. In this local authority area, one third of children live in poverty.

Tom arrives promptly in a large silver 4x4. He's just as I remember him: friendly, down to earth, and highly personable. My first encounter with Tom came at Edinburgh Castle, in a grand hall which is not open to the public, where I addressed an audience of politicians and corporate leaders at the launch of the Scottish Television Children's Appeal – also sponsored by Tom.

He takes his place in *my* seat in the big warm car and I humbly climb into the back of the little cold one. We set off and drive for a minute before pulling up at the last house on the only road out of town. The white detached dwelling, set in two floors, is where Tom made his beginnings. His father was a local grocer who faced financial ruin when Thatcher closed the pits. He learned the basics of shop-keeping by studying his father's stewardship of the family business, his love and respect for his father making it all the easier to absorb his teachings.

As deindustrialisation took hold, unemployment rose sharply. The UK government made grants available to people looking to start a business. Tom leapt at the opportunity and, in the economic maelstrom, cobbled together for himself a modest home video delivery service. The business model was simple – people contacted him by phone and requested a film, and sometimes even an alcoholic accompaniment which Tom acquired and promptly delivered. By purchasing large quantities of vodka, he had only to mark up the price slightly to turn a tidy re-investable profit. It was when he applied this age-old principle to selling sports shoes that his hard work and a government grant (and a £5,000 loan from his dad) paid off. In 1998, Tom sold his Sports Division empire for £300 million to quite the fanfare. Celebrations were had but soon it came time for Tom and his wife to decide what to do with all their money.

'Well, I guess what philanthropy means to me, Darren, is that you see where I was brought up,' says Tom, gesturing out to his stomping grounds. 'I was very lucky to come into a great deal of money, and my wife and I had a couple of big decisions to make. The first decision we made was we don't want to be the richest folk in the graveyard.'

This is a killer line. Something all high-profile philanthropists must possess in the media age. I get the impression Tom has dropped this gem before. His rehearsed recital of his early life and dizzying social ascent is a party trick I recognise, as it's also one of mine. People who 'beat the odds' and 'make something of themselves' rarely forget it – they aren't allowed to. The lowborn who go onto achieve success must provide context and reasoning for the benefit of those of higher social castes as to why a working-class person walks among them. This becomes your 'story', which is then integrated into a heavily revised narrative. You recount it for the benefit of others but if you're not too careful, you may start believing it yourself. It doesn't quite reflect the whole truth – it tells the 'audience' what it wants (and needs) to hear. Tom's story is compelling not simply because it is extraordinary but also because he's told it so many times; he knows how to steer it home. Familiar as I am with Tom's tale, and why he tells it like he does, I am still somewhat transfixed.

'We wanted to do something with our money when we were still here. That was something I learnt from Carnegie. We wanted to try and make a difference for folk.' Hunter has, in many ways, modelled himself keenly on the Dunfermline-born steel magnate who, like Tom, began working under his father. After the industrial revolution left their textile business on the scrapheap, the Carnegies borrowed the price of their fare across the Atlantic from a local baker and fled the UK to the land of opportunity. Prodigal son Andrew rose the ranks of the US steel industry, eventually becoming the world's wealthiest man and, arguably, the most culturally ubiquitous industrialist in history. His story remains a cornerstone of a powerful and intoxicating capitalist mythos: anyone can achieve whatever they set their minds to if they work hard.

'You can't do it for people,' Tom insists, 'but you can help create the environment where they believe there's a way out of poverty, there's a way out of where they are, there's a better way, they can start a business.'

Hunter speaks with a passion and certitude which is invigorating. It's no surprise that, with his impeccable communication skills, relevant lived experience and eye-watering riches, successive Scottish governments have made countless overtures to him; in an age of deregulated capitalism, Hunter stands as a living exhibit, in the misty eyes of free-market fundamentalists, that anyone can 'make it'. For most people, this is a lie, but one which persists precisely because the few who do succeed are held up as the rule and not the exceptions to it. Tom is by no means ordinary, nor is he being coy. It's clear he believes in the transformative power of enterprise. And why shouldn't he when that has been his experience? 'They can do it,' he insists, 'a belief that something better is down the line.'

Despite his success, Hunter remains in admirable proximity to his former community, which seems to hold him in high regard. Our interview, taking place by a busy roadside, is constantly interrupted by the honking of car horns and locals gesturing or stopping to say hello. One woman even emerges from her home and hands Tom some old photographs. This is something I always remembered about Tom, from years back when he came to visit a community project I used to run in Glasgow's Ibrox. He isn't driven around. He doesn't have a team of handlers buzzing like bees around him, holding doors, taking notes, fetching things. And despite his fortune, he and his wife believe their offspring must make their own way. While their margins for error will not be as slight as most, Tom's decision to give his money away when he dies must surely have raised a few eyebrows at the dinner table over the years.

Hunter is a rich individual who a washed-up, pampered leftist with radical pretensions should never meet. When confronted by a man who appears thoughtful, sensitive and kind, as well as loved and respected by working people, who just happens to be worth a few quid, it places the old ideological framework under immense strain. In the presence of an everyman who has grown wealthy by working pragmatically within the parameters of society (rather than trying to change it), and who possesses a compelling rationale which accounts for his success, the Marxist refrains that roll so easily off tongue from the comfort of the armchair suddenly seem ill-fitting and self-righteous – much to my frustration.

We jump back in the cars, ascending the hill up to what looks and feels like the more 'deprived' area of New Cumnock. We disembark from our respective vehicles and Tom takes a phone-call while the director, producer and I discuss the sequence we will imminently film. A young man, keen to make his presence felt, approaches and then encircles us once on a motorcycle, which roars and startles me, before he speeds off.

Ordinarily, we ensure Covid-19 safety guidelines are met by using a two-metre length of tape which is attached to my ankle and then the ankle of the contributor, thus marking the required distance between us. But this morning, the team and I agree to relax the requirement for Tom. We have no reason to suspect he would object. Tom has not asked for special treatment. Yet we bend the rules for him anyway without second thought. Given our films are about class, the decision is all the more perverse.

Tom and I cross the road and meet at the top of the path, overlooking New Cumnock. Despite the sunshine, it is bitterly cold and both Tom and I realise we are under-dressed for this altitude. We step onto the path and begin the walk down before I am stopped and informed that we are still too close to one another. 'You'll have to walk on the grass, Darren,' I'm unceremoniously instructed by my director. I experience that pang of fury once more. Why must I walk on the grass? This is my television series! Tempering my inner diva, I swallow down my indignation, plunging my red limited edition Pumas into the thick, brown mud, by now thoroughly aware of my place in the pecking order. We resume our walk and talk, but the conversation does not come as freely as before. The sequence is supposed to be about philanthropy but other questions are forcing their way stubbornly to the forefront of my mind. While agreeing philanthropy has its place, it is also highly problematic. I put to Tom some of my concerns, hesitantly, battling the strange sense of deference I feel in his presence.

'Yeah, I mean, when we invest and we invest with government quite a lot, but we'll always say we are no substitute for the taxpayer. So, we want the government totally committed to it,' he tells me, referring to the various social enterprises his foundation supports. 'And we agree what success looks like up front. And we will take the risk because governments don't like to fail. And if we prove it, we say you, the government, has got to adopt it as policy. So that's all agreed up front.'

I'm fascinated by Tom's conception of government. One clearly shaped by his wealth. For Tom, the state represents just another market force with which he can negotiate. For him, government is not the ultimate authority in society – it's a competitor and a partner.

Tom really comes to life when discussing the change he envisions, the young people he would like to encourage and the businesses and social enterprises he'd like to empower. His passion is undeniable and he has a track record unparalleled in modern Scotland. Tom's Hunter Foundation has so far given away £100 million to education and entrepreneurial causes that seek to, in Tom's words, 'positively disrupt society'. It is, perhaps, quite typical that I harbour a mischievous desire to – even just for a moment – positively disrupt Tom. Pampered as I have become and blunted as my senses are by my recent affluence, a raging red fire still burns within me that occasionally makes an impromptu appearance – whether I like it or not. As great a storyteller as Tom is, I cannot banish from my mind the correlation between extreme wealth and exploitation. Nor can I unknow the fact our tax and legal systems are configured, often in partnership with the wealthy or their representatives, to preserve the myriad advantages enjoyed by the rich. Despite Thatcher's hope that the great wealth amassed by economic liberalisation would be spread around, many in Tom's former locale are clearly still waiting for some of it to trickle down.

I venture into terrain I sense Tom will be less comfortable or rehearsed in discussing, gently probing him on the influence of the wealthy on politics. 'Because you've got a big cheque book, doesn't mean you're right,' Tom says, perhaps another killer line, though deployed with slightly less surety than his previous statements. 'And the way political parties are funded is fundamentally wrong, in my opinion. But what I can do with the system that's there just now until it changes – come the revolution, Darren – we have got a wee bit of influence and we can give people who know what they're doing a platform from which to be heard.' Tom senses where I am going. He's aware of my true political leanings, despite their becoming slightly muted in recent years due to the fact I now have something to lose. His jovial reference to 'the revolution' is proof of that.

'You keep bringing me back to class, Darren,' Tom observes, correctly. I put it to him that his foundation, which combines the cultural and financial capital that comes from great wealth, and the imagination, innovation and vision of those less fortunate, is an example of interclass collaboration. The purpose of my question is to hear Tom acknowledge the existence of class as an objective socioeconomic fact.

Despite being nearly two hours into our discussion, Tom has shown little interest in my countless invitations to explore class at any depth. One may assume this is because, in acknowledging the existence of class, his rags-to-riches story may be re-contextualised in a manner less flattering to him. There is, after all, an alternative version of his hero Carnegie's achievements in business. One which involves poor pay and working conditions, favourable treatment by governments who perhaps should have reined him in and the infamous Homestead Strike in 1892, where industrial action over a pay cut resulted in workers of the Carnegie Steel Company being callously gunned down. Acknowledging class makes trickier terrain for the rich. And, like my armchair Marxist refrains, places their capitalist ideology under immense duress.

But what I sense is that Tom's reluctance to discuss class is not necessarily a denial of its existence, nor motivated by a desire to paint himself in a charitable light, but, curiously, it's rooted in a belief that by ascribing a social class to an individual, you make an unfair value judgement about them. It is the attitude that by identifying yourself as a member of a social class, you internalise a self-limiting belief. From this cultural vantage point, class, whether adopted or ascribed, is seen as reductive; erasing the richness and complexity of the individual in a manner which says little about their values, their abilities or their experiences.

'I never judge a person by, you know, their class. I hope I don't,' says Tom, regaining a foothold in the conversation. 'I judge them by what they've got to say, what their attitude to life is – are they a doer, are they going to make things happen, are they going to change things for the better from whatever background?' I sense I am now speaking with the real Tom Hunter, and not the famous, media-savvy philanthropist. For a brief moment, we are equals, as he locates, behind a wall of well-worn soundbites, what he really thinks and believes – not about philanthropy, but about class.

'I've never been held back by class; I never saw it as a barrier. I think we're all different but you've got to have the belief that you can get out of the circumstances you're in, it's as simple as that.' Tom is correct in a sense: an individual's inherent ability, attitude, work ethic and resilience have a significant bearing on how they will rise to meet the challenges associated with their circumstances. But an individual has no control over the circumstances into which they are born. And it is those circumstances which, in many cases, will determine the trajectory, quality and even length of someone's life. Inequality will always, to some extent or other, exist, but the persistently extreme nature we see currently is quite simply a moral and economic outrage, given the wealth generated for a tiny fraction of the population.

Tom is just one of many famous sons and daughters to hail from New Cumnock. But what is most interesting about the town, to me at least, is also something its local authority is not quite as keen to parade as the outdoor pool or whatever pub Robert Burns allegedly once had a pint in. East Ayrshire is greatly over-represented in Scotland's prison population. For more than 20 years, more than half of Scotland's prisoners have come from the same deprived areas. Many of those kids, born into poverty which predisposes them to social exclusion and thus prison, will not possess the skills and attributes philanthropists are often looking for. They won't be 'go-getters'. They won't have 'good attitudes'. Who is going to give them a hand up? At its core, philanthropy is a strange form of benevolence rooted in a belief that the rich know best as to how their wealth should be redistributed. That they are supremely placed and endowed with unique foresight which eludes the state. I do not doubt some of them are. Nor do I deny many wealthy people genuinely wish to make a positive difference and indeed, do so. But while Carnegie may have built a lot of libraries for the working classes and publicly supported the trade unions that fought for them, privately he often acted against their interests – to advance his own.

Despite the interview going well, I can't help feeling this morning was an opportunity squandered. As we go our separate ways, I begin to kick myself that I did not put to Tom the many other questions I had in mind. Some based on what I already know and some on what I have learned throughout the course of my work on class. For some reason, when it came to putting tougher questions to Tom, words failed me. They failed me for the same reason I was placed in the little car and why I grudgingly planted my pristine footwear in the muck. They failed me because in my bones, and to my utter consternation, the presence of great wealth, prestige and power elicited in me an uncharacteristic and undue deference. Intuitively, I knew, as did my team, that Tom is accustomed to a certain quality of conversation. A certain quality of treatment. And a certain quality of questioning.

How we understood that without it ever being spoken or even acknowledged may seem mysterious. Why Tom never stepped on the grass and into the mud may appear inconsequential. But you will find a similar deference in political and royal correspondents. In how we fawn over celebrities. You'll find it in the classroom and the court-room and on the shopfloor, to varying degrees. And you'll find it in elected officials when they deal with people like Tom. A reflexive def-erence based not necessarily on someone's superior knowledge or insight, such as when we defer to our doctor on health matters or to a pilot when jetting off on holiday, but far more primal in nature. This 'instinct' is perhaps the subtlest, most decisive advantage enjoyed by those of higher social classes. Only when you sense it in yourself, des-pite your radical pretensions, do you come to realise why a politician might struggle to speak plainly to the CEO of a tax-avoiding corpor-ation, or a member of the royal family, or land and property owners with respect to thorny issues of fairness, justice and equality. There is often more than simply an alignment of interests at play in the rela-tionship between wealth and politics, but also the near irresistible urge on the part of our elected representatives to defer to their betters – the super-rich – in much the same way the general public often defer to the political classes – for no good reason whatsoever except an unconscious reading of social rank.

Just as distrust of authority can become pathological, causing some to grow unnecessarily hostile towards health advice or police, creat-ing discord in poorer communities, so too at the other end, misplaced faith in the wisdom, benevolence and competence of wealth and priv-ilege can be just as self-defeating – and even more dangerous. I wouldn't lump Tom in with the ruling classes. And while aware I will score no brownie points from my comrades on the left for saying so, my feeling remains that Tom is a product of his environment rather than the environment being a product of him – a subtle but important distinction between those who just happened to become rich and a dyed in the wool ruling class. Still, while certainly an elusive aspect of any class-based social system, misplaced deference plays a significant cultural role in preserving the imbalanced economic status quo which, in my view, characterises British society.

What was most illuminating in my time with Tom was his response when I asked if he had ever issued a similar rallying call to his wealthy peers about their interference in politics as he has done consistently to government officials. His reply was that he 'didn't like to preach' – perhaps a subtle and unconscious form of deference to those fortunate enough to be even wealthier than him.

II

Speak Properly
Talking your way in and out of poverty

To me, words are like music. When arranged in a particular way, and written or spoken with a certain conviction, an alluring harmony is produced which I find immediately arresting. What is being said, it's meaning or, indeed, whether I agree or not comes entirely secondary to this initial capture of my fleeting attention. I am often propelled by a sudden, ferocious interest into a particular field of thought or study – not necessarily by a desire to educate myself on a specific topic, but because I am drawn to how someone writes or talks about it. Much like a great tune, which can be enjoyed without any real understanding or foreknowledge of its genre or era, well-arranged words, expressing fluent, coherent ideas, are simply music to my ears. And, to stick with the music analogy, it shouldn't matter if the material originated in the mind of an Oxbridge graduate or a guitarist who learned their trade on the dole: if they can play, they can play.

My lifelong fascination with language and the subsequent capacity I have developed for speaking is not something I consider remarkable. Yet, as I've moved out of hardship and into cultural and social spheres which are dominated by the middle classes, I am increasingly aware of how surprised people are when they hear someone from a working-class background express themselves with a degree of articulacy. As a 'diamond in the rough', currently ascending the social scale, I encounter people from higher social classes more frequently.

I recall one Scottish book festival where I arrived and was looked at like I was there to collect the bins. Often, touring the country, I feel like a living art installation that middle-class people pay money to interact with. As I attend more events and engage in more media, I get asked more questions. Some are thoughtful. Others are personal. And some of them are downright rude. Irrespective of the quality of the question, or my enthusiasm to address it, a great deal of my time is now devoted to furnishing my various inquisitors, on social media, television, radio and even in the street, with polite and satisfactory responses. The question I least enjoy answering is also the one I am asked most frequently: 'Where did you learn to speak so well, Darren?' The people who ask me this question always think they are the first person to ask it. Countless journalists, public officials and book-festival enthusiasts quite simply cannot restrain themselves. They don't even realise how insulting it is to be asked such a question. Their aloof enquiry is made so earnestly that rather than take offence – which would be well within my rights – I have learned to contort myself to grudgingly accommodate it. What these people are really broadcasting is that they are somewhat surprised by my ability, as a working-class person, to string a coherent sentence together without soiling myself.

I have since developed a standard response to this question: a paraphrased, conversational version of the 'words are like music' passage you just read. I have adopted that as my go-to reply because it's a lot easier for everyone involved if I don't say something like: 'Why shouldn't I be able to express myself clearly? These are my words, too – middle-class cunt.'

My more palatable response meets all the necessary criteria for successfully engaging with a middle-class person, thus hopefully evading any potential power-play in which they may engage if I were to upset them. Firstly, it does not challenge them. Secondly, it is not confrontational. And lastly, it educates them in a way that does not make them feel like they do not already know everything.

When dealing with many middle-class people, these are skills you come to depend on if you come from my kind of background. But it is exhausting. In my career as a musician, writer and public speaker, I have performed countless hours of emotional labour in an effort to make it easier for people who regard themselves as informed, cultured and sophisticated to advertise their ignorance by insulting me. Being polite is important, of course. But even the virtue of good manners is part of the dance. Moving up the food chain, and out of poverty, often requires more than just talent, competence, determination and good fortune – you need to be willing to shut up at precisely the right moment.

Your social mobility depends as much on your communication style as it does on your grades. Indeed, in some cases, how you come across may be of greater import than what you are actually capable of. I recall a visit to a community space in Glasgow when lockdown restrictions were eased for the first time. I had been invited to give an 'inspirational' talk to young men who were at risk of being drawn into the criminal justice system. Prior to arrival, I decided to devise a short workshop instead. My aim was to test a theory I had been pondering about self-limiting language and the attitudes, values and behaviours that may arise from it.

Most people take their ability to express themselves for granted. Even those who regard themselves as introverted possess a baseline ability to communicate effectively when they must. But those who grow up in hostile environments, deprived of support and encouragement, tend to have only a few hundred words at their disposal by the age of 16 – not because they are stupid but because these words cocoon them from vulnerability. I explained to the boys that when I was growing up, I felt a lot of social pressure to speak a certain way. To present a certain demeanour. That my failure or inability to do so would result in derision and sometimes even violence. I then presented them with a list of words, with varying emotional resonance, and asked them simply to raise their hands to those that they would be prepared to say in a regular conversation.

Hopeful
Stab
Beautiful
Ragin'
Eccies (ecstasy)
Love
Hate
Romantic
Scheme
Grateful
Considerate

Buzzin'

Community

Tuberculosis

Moist

Book

Bucky (Buckfast tonic wine)

The list was specifically designed to draw out why some words were deemed more appropriate than others. Naturally, those words which conveyed aggression rolled easier off the tongue. Some words provoked laughter while others created an uneasy silence. I explained to them that by limiting the range of words at their disposal, they were effectively imprisoning themselves. They were, out of a desire to portray a particular tough image, inadvertently impairing their ability to develop as men. I then set them a task. I would go round the room and each of them had to describe one memory where they felt happy, connected and supported. The catch being they were not allowed to swear and had to describe their emotions as accurately as possible. It's in these moments that class clowns retreat from the spotlight and hardmen go limp.

'Eh, I remember this time I fuckin—'

'No, stop. Go back. No swearing.'

'I remember this time I did an apprenticeship. I got a job in a factory. It was heavy.'

'OK. Good. What do you mean by "heavy"?'

'Eh, it was heavy. Heavy good feeling.'

'OK. What was that feeling?'

He looked around the room as his peers awaited the response. He looked at me and, then, that vulnerability surfaced.

'I was really happy,' he said. His eyes misty, his body language open, his entire demeanour shifted.

I invited the room to applaud. The air was electric. The feeling among the group was one of elation because everyone understood what a tremendous leap of faith he took simply to express his emotions in a clear and honest manner. The other boys, with some trepidation, embarked on the same process. One recounted the time he built a shelf and how proud he was that he saw it through without giving up in frustration. Another told how he 'loves playing and watching football' because he 'forgets everything else'. And the last boy shared the memory of holding his baby brother for the first time and noticing how tiny and vulnerable the newborn felt in his arms.

Each recalled an instance where they felt connected to something larger than themselves, from which they could derive a sense of meaning and purpose. The factory. The football fields. The craftsmanship. The children. Traditional pursuits young men not so long ago would have been thrust into by the natural flow of a bustling community life. Today, communities do not bustle in quite the same way. No longer organised around pits, shipyards, mills or forges, they are now situated around distant abstractions, like consumerism and property ownership, which remain out of reach. The factories are gone. The fields are strewn with broken bottles hidden by unkept grass. The tools have been downed. The families broken. The social bonds which once adjoined working people to a shared experience have been torn, leaving many youngsters rudderless, desperate for attachment. Where there is no wider purpose, and no clear sight of their future, these boys instead attach to one other and their shared sense of exclusion in acts of unconscious commiseration and ritualistic self-sabotage. In this context, restrictive language and oppressive speech become connective tissue which provides a cultural perimeter around them, thus binding them in their uncertainty.

Despite my love of language, and my ability to wield it confidently and expressively when I have to (and sometimes when I don't), I still run into problems, occasionally – not due to an inability to communicate but because of the expectations of others with respect to what people like me are capable of. In 2020, I appeared on *Question Time* – from the comfort of my living room. It was my second time on the show, having lost my *QT* virginity to the charming David Dimbleby a year prior. I was as relaxed as it is possible to be ahead of such an event. Sat on my couch, my home comforts around me – and a visible plethora of books I had no intention of reading neatly shelved behind me – I was overcome by a calming sense that all was in hand.

Unbeknownst to me, however, were minor details that would coalesce throughout the broadcast, resulting in my most disastrous television appearance yet. The first was a four-second delay between myself and the guests in the studio. The second was that I was unable to see anyone. And the third (which I learned having glanced at Twitter while the other panellists were talking, to find hundreds of angry tweeters complaining that I was 'cutting over' other guests) was that we were being broadcast live – not filmed an hour before, as was usually the case. Hilariously, I made this discovery when presenter Fiona Bruce put a question to me. I had no choice but to come clean with her. 'I've just been told we are actually live,' I laughed nervously and, in doing so, created just enough time to gather my composure and address the question which, coincidentally, was about inequality. I cannot recall exactly what I said but I do remember speculating that developed economies like the UK and United States may have been experiencing high Covid mortality rates due to their respective places at the top of the social inequality table. That there might be some link between class inequality and the rapid spread of the disease. It was hardly scientific but that was my pitch. Little did I know that this was going down terribly online.

'Who is this communist?' wrote one temperate gentleman. 'Britain is the most unequal country in the world? What about Africa?' wrote another geography expert. The time delay wasn't helping, nor was my being unable to see the panel. The confusion culminated in none other than then *Good Morning Britain* host Piers Morgan tweeting out a hastily snapped screenshot, in which my cranium appeared hideously enlarged, with the following, characteristically fair-minded caption: 'Who is this raging lunatic?' After that, I don't remember very much except that sinking feeling that engulfs you upon the realisation something has gone very badly.

Like a true sadist, I trawled the hundreds of replies to Morgan's provocation. Many of them were supportive of me. A lot of people across the UK clearly understood there were technical problems and that I was not being intentionally rude. But others, prompted by Morgan's framing of my demeanour as angry, piled on with their own stunning observations. Some speculated that perhaps my years of alcohol and drug use had impaired my ability to communicate. Others questioned why I was not subtitled. It was clear that despite my not exhibiting even a hint of aggression throughout the 59-minute broadcast, as well as the great pains I went to in order to express myself clearly, aware I was addressing a mass audience, a significant number of viewers still regarded me as some kind of illiterate drunk who had somehow wandered onto a national television broadcast.

The fact that my grasp of English was as firm as the two politicians and the scientist also on the panel appeared to elude many. Thankfully, this precarious cultural terrain was by then already very familiar to me. Indeed, my speaking style is, itself, a direct consequence of operating across a variety of social and professional settings, where the expectations and demands of communication vary wildly. An average day might see me converse with a man like Tom Hunter, before meeting with some young offenders, speaking at a political or corporate event, and then attending a recovery group ahead of a national television appearance. It is not possible to traverse this cultural terrain effectively (to the extent that I do) unless you have developed something of a repertoire. The difference between me and the people who are most critical and judgemental of my speech style is that my critics tend only to mix in one social sphere. The irony here is that people like me, who can comfortably modulate our speech around extensive vocabularies and retrain our ears on how others speak (to graciously accommodate their inferior communication skills), are often judged harshly by those whose communicative ability is rarely tested. It is assumed that those from loftier social classes are the most adept communicators but this assumption remains theoretical, given they spend most of their lives chattering among themselves.

Language and speech are like currency. We use them to trade knowledge, information and sentiment. Some trades are more exacting than others, not least where vast sociocultural and economic distances are concerned. The dominant middle-class speech style in Britain, like the dominant global currency, is not a reflection of any inherent attributes or utility – it's an expression of what is most convenient to the dominant social classes. The United States dollar is traded globally not because America is the best, smartest or most sophisticated country in the world, but because it's the wealthiest. In Britain, the prevailing speech style – referred to as received pronunciation (RP) – associated with social respectability is not particularly sophisticated or efficient but rather acts as a broadcaster of rank. It's dynamic, not because it requires any specialist knowledge to deploy but rather because it transmits on two frequencies simultaneously: as well as what is being said, this speech style discharges subtle affectations, imbuing the speaker with a veneer of respectability, trustworthiness and sophistication – whether it's deserved or not. This voice, which is difficult to describe but instantly recognisable, is one which most people from the lower orders will come to associate with authority. It's regarded as universal because it ports so effectively to every social setting in Britain. Everyone understands it. But it's just a bunch of words and sounds like every other speech style. It simply becomes 'superior' because it is used by the middle and upper classes, much like property gains value merely by being occupied by them. And unlike most other speech styles, it does not emerge naturally from a geographical location but is, instead, a highly standardised manner of speech arising primarily from institutional conditioning.

I've been studying the middle classes up close since I was a teenager. Not because I found them especially fascinating but because there was no way around them. If you wanted to get on in life, you had to learn the steps of their dance. You had to remain in their good graces or life could quickly become difficult. I understood, early on, that middle-class people held the keys to whichever doors I may one day wish to pass through. This process of learning began, as it often does, in the classroom, where often, the toughest teachers spoke in this prevailing authoritative voice.

I recall one episode when a teacher forced me to admit to starting a fight in the playground, even though I hadn't. She pulled me out in front of everyone and repeated her question: 'Did you start the fight with Peter?' To which I kept telling her 'No.' She reduced me to tears, in full view of two open-plan classrooms. After a humiliating onslaught which lasted for ten-or-so minutes, I simply conceded her groundless accusation. That day, I learned something about challenging that voice.

It was as evident to me then as it is now that she was not as interested in the cause of the altercation as she was in her incorrect presumption being validated in front of an audience. She couldn't let go of her false belief that the fight had been my fault. And in the classroom dynamic which she dominated, she was not obligated to contend with her mistake. For I was just a child, one she terrified to make a powerful example to other children that retribution awaited any who dared to challenge her. To this day, I'm grateful the belt had by then been outlawed. Judging by how she slammed her fist down on the desk when I 'admitted' to being the culprit, I suspect she'd have taken great pleasure in teaching me a painful lesson.

This dynamic was echoed at various stages of my life in systems of secondary education, as well as health and criminal justice. As with my old teacher, it was about more than accent or rhetoric. It was about conveying that the conversation occurring was a one-way street – you are the recipient of instruction and failure to respond accordingly is punishable. Hierarchies are inevitable and authority is often necessary, as is deference in certain circumstances, but problems can arise where gaps in power and accountability are found. Think back to the drug crisis, and the ravine between those with lived experience and those enmeshed in the drug sector. Think of the man in Possil who decided to remain at home despite having a heart attack because he felt the doctors at the hospital wouldn't believe him.

When people with authority are hard to hold to account, their presumptions go unchallenged and, over time, set like cement. Like their voices, their bad assumptions are integrated into a sensibility which prevails as 'common sense', irrespective of its basis in reality. They may begin to lose touch with the possibility that they, like all people, possess the capacity to be wrong, to be aggressive or to behave vindictively. As a consequence of the reverence with which others are conditioned to treat them – and which they themselves may come to expect – poor behaviour and malpractice often go undetected by both parties on opposite ends of the power dynamic. And in the rare event their assumptions are challenged or contradicted by those of a lower social standing, who often speak with an 'improper' voice, the polite and temperate veneer so central to middle-class studied informality acts as a cultural decoy, masking passive aggression and power-plays, and concealing improper behaviours and attitudes rarely associated with people who speak as they do.

'Language attitudes research shows that as soon as we hear somebody, we make social judgements about them,' says Professor Jane Stuart-Smith, who teaches phonetics and sociolinguistics at Glasgow

University, in a Radio 4 English accent which, to my mind, places her firmly in Edinburgh's middle class. 'And those social judgements have got nothing to do with what they're saying or how they say it,' she adds, 'but rather it's the connection between certain speech styles, certain sounds, certain words and certain parts of language with particular stereotypes.' She tells me about a language attitudes study conducted entirely over the telephone. 'This was an ethnicity study and just from the word "hello", people made their judgements on who that person was, were they trustworthy, what race they were, and so on.' She then draws a parallel between widespread attitudes around ethnic accents and those of young people from working-class backgrounds. 'So, I think it's particularly difficult for young adolescents who come from deprived areas who speak in a strong dialect. If there's any kind of telephone interview, literally the first few words and the job's gone. And it's got nothing to do with them and everything to do with the accent.'

Might this also occur in the arts? My experience tells me, absolutely. In 2017, I attended a Burns supper in my capacity as a rapper – which is just a poet for black and poor people. I say 'attended', what I really mean is gatecrashed. At Burns nights, where Scotland's national bard is celebrated, middle-class audiences enjoy the thrill of subversive, challenging, vulgar, edgy political poetry – as long as it's over 200 years old. I have become accustomed to the awkwardness that descends on a room full of people who think they are interested in literature but who seem to have acquired a rather narrow view of what it may entail.

The thing about performing at events attended mainly by middle-class people is that they are supremely confident in their insights. At any moment throughout the course of your performance, they may feel entitled to raise their hand and ask you a question because, for them, life is just one big TED Talk. Part of the piece I performed is featured below:

'How come its bright in this posh park but in Pollok its dark, prolly coz the sun isny shining oot everybodies arse
   vegan liberals lecturing me to buy local while they're sneering at the glasgow dialect in my vocals'

After I finished, while absorbing the rather confused reaction of many in the audience, who likely were expecting some twee interpretation of 'To a Mountain Daisy', a young woman raised her hand. I became immediately anxious. Would she challenge my use of profanity? Or perhaps attribute the misogyny of the central character to me, personally? Had I mispronounced 'renaissance'? I didn't want to engage with her through fear that the show may become bogged down (like on so many previous occasions) by a middle-class audience breaking into a conversation with itself.

In situations such as these, you must act quickly and with confidence. I decided to acknowledge her, pointing her out to the crowd, before inviting her to pass comment. 'Did you really live in a bin?' she asked, referring to a throwaway line in the final verse, delivered tongue-in-cheek. The only line in the 48-line piece, performed from memory, that I did not rewrite 30 times. Throughout the course of a three-minute performance, where I had deployed every salvo in my artistic arsenal, describing vividly the socioeconomic, cultural and political forces that bear down on those regarded as 'lower class', that was all she had to say? And there we were, in the same room, in the same city, both hailing from the same country, yet separated by a vast cultural ravine.

I relay some of my experiences to Jane, noticing that her body language has shifted throughout the course of the interview. She is more open than she was when she arrived. Less nervous and inhibited. Often, when I meet middle-class professionals who are aware of my work, they adopt an apologetic demeanour and occasionally engage in rehearsed self-deprecation. A performance of deference I assume they think may be disarming. They do this because they assume, on some level, that I am out to get them. That I am angry at them. Countless professionals have cracked jokes about their big cars and their posh communities. Of course, these defences dissipate the longer we spend in one another's company. A natural rapport is generated merely by occupying the same physical space and interacting over a period. Multiple proximity gaps can be reduced simultaneously simply by spending time with each other. Due, however, to Britain's structure as a class system, people from different social classes rarely interact and when they do, the interaction occurs around a power dynamic which elicits unnatural behaviour and speech.

In my experience, this rapport can only be generated when common ground is found. But before this can occur, we each have a number of presumptions, stereotypes, judgements and insecurities to sort through. We must get a measure of one another in order that we may better orient ourselves within the interaction; the way we speak, perhaps more than anything else, is the first obstacle which must be traversed.

Accent Bias Britain, a project dedicated to addressing the role accent plays as an indicator of social and ethnic background in the UK, defines the prevailing voice in Britain, received pronunciation, as 'the accent of English in England that is associated with people from the upper- and upper-middle-classes'. RP was widespread among fee-paying public pupils and universities by the end of the nineteenth century, though RP's popularity has declined since then and is believed to be the native accent of just 3 per cent of the UK population. Despite its dwindling prevalence, RP remains the national standard and is believed by many to be the most prestigious accent found in British English. In Scotland, during the industrial period, countless middle-class families believed speech to be such an important marker of social class that they sought the aid of elocution lessons to rein in their regional accents. This is partly why so many people in Scotland are mistaken as English (and often receive abuse as a result) when they are, in fact, Scottish.

Of course, Jane lives in Edinburgh and works in Glasgow. She isn't upper class. Indeed, I couldn't even hazard a guess as to whether she is English and for the purposes of this argument I have not checked. The point is: Jane's accent is, in my view, part of a performance. It's a way of speaking which she has developed, one that supplements her native accent, which she intuitively presents when she desires to be taken seriously. The young men in Pilton do the same, except they modulate their speech and vocabulary to convey toughness and insensitivity, thus warding off potential threats of violence. Tom Hunter's speech style remains consistent irrespective of who he talks to – one of the perks of being rich. Like Jane, I too am engaged in somewhat of a performance, but while my speaking slightly more 'proper' might strike someone from my housing estate as pretentious and fake, for someone of Jane's social caste, my working-class heritage is a truth I simply cannot conceal.

'I'm lucky,' admits Jane with admirable self-awareness, 'because studies have shown that if I was to say something in my accent, people will believe it.' Having not read these studies, I decide to take her word for it. As she speaks, I try to imagine her relaying this research in a Geordie accent, perhaps sporting a fresh facial scar. 'Whereas if you were to say exactly the same thing in your accent, they wouldn't believe it. And this goes down to the connections between language and social stereotyping and ideologies. There's nothing special about my language.'

Oh, but there is. The British Council, the UK's international organisation for cultural relations and educational opportunities, notes in an article titled 'What does your accent say about you?' that 'Within the UK, people have prejudices about the inner-city accents of Birmingham, Glasgow and Liverpool, which place these accents at the bottom of a league table of approvals,' adding, 'but this situation is changing.' Though evidently not fast enough if my *Question Time* appearance was anything to go by. A shift in language attitudes towards regional speech styles requires greater exposure to them, through media. But hardened attitudes to regional accents create resistance among media executives and producers to broaden out what is seen and heard across programming. I once caught the tail-end of an email chain I was not supposed to see, where I discovered that the reason I had not been offered television work at UK-wide level was because producers at a reputable broadcaster feared they would have to subtitle me – that is how daft audiences are assumed to be.

Jane explains how the development of vocal anatomy often constrains a person's ability to truly modify how they speak, even if they wish to – not the news I was hoping for given my ambitions in broadcasting. 'From the very beginning of learning how to speak a language, we also learn how to hold our vocal tract in a particular position. So, when somebody speaks central Scots, they have learnt not just the words and the segments and the intonation, but also learnt how to speak it. I've learnt how to speak my variety, which is, you know, it's a bit breathy and a bit of a lower larynx.' She kindly demonstrates how these divergences in how vocal anatomies develop between social classes by doing an impression of me, mimicking my pronunciation of the 'r' sound, which excruciates me – because it is so accurate. Hearing my accent erupt from the mouth of a posh English woman is extremely unpleasant. Why would she vandalise her vocal cords in this way? The sound rouses in me a deep sense of shame. The way I think I sound when I speak is not how I really sound. Conversely, Jane does not pronounce the 'r' when she speaks naturally, instead producing a rather whispery elongated vowel sound I can only describe as 'argh' which is, for some reason, easier on the ear than my version.

'I lecture to a very large cohort of students every year,' she tells me. 'I have to talk to them about social class and language. And every year I get students writing to me angry that I've talked about it and also angry that I'm even talking about the idea that Glaswegian vernacular is a valid language to study.' Jane is describing the unconscious value judgement many people make of dialects which are regarded as lower class and, by extension, unsophisticated. 'And I have to turn around and say, "Have you ever tried to speak Glaswegian?' It's very difficult!"' she exclaims, exuding a deep respect and wonder for my accent that is clearly lost on her middle-class students. 'They are angry because they've come through a system which has given them the feeling that there are certain varieties of language which are worthy of study and others which are not.'

I experience a strange mix of validation, resignation and fury as the subtler contours of class conflict are outlined in such vivid scientific detail. There is the validation that I wasn't wrong to have assumed that my accent marked me out in the eyes of others – not just on *Question Time* but throughout my life. The resignation that the fact my social range exceeds that of many middle- and upper-class people, who wrongly regard me as lesser, does little to temper their unearned sense of superiority. And then the fury upon the realisation that if I, as an individual, must still confront these barriers despite being fully aware that they exist, as well as being publicly noted for my work, how do those with no insight and no cultural capital to fall back on navigate such a treacherous, punishing cultural landscape?

Rather than liberating, this revelation is dispiriting. For it means I often communicate that which runs entirely contrary to what I intend as it is encoded deep within the sub-structure of my speech. When I speak, I do not simply betray that I am lowborn, thus eliciting in many the assumptions that go with that, but whatever I say, no matter how precise or inspired, will always be interpreted by some as the furious ranting of a junky with wet-brain.

Just as language and speech often arise from class-specific contexts, so too do certain customs, beliefs and values. That Jane's middle-class students are often so dismissive of the Glaswegian dialect says more than I ever could about cultural inequality. Her students, and the tens of thousands of other students attending prestigious British universities who harbour this unconscious disregard of working-class culture, will ascend to influential roles in education, the arts, media, politics and criminal justice. They will ascend, in part, because of their bias. Bias which can only be corrected by being exposed to the great variety of speech styles which exist beyond their own standardised speech. Having ascended, they will, from their ivory towers, exacerbate the very inequalities which they, themselves, benefited from, while claiming to be trying to reduce them. When their grand plans and schemes come crashing down on the rocks of social reality, they will fashion new myths and false beliefs to account for their failures. And as long as they speak 'properly', few will ever suspect that they, in all their wisdom, could be the root of the problem.

The wrong accent may conceal true insight and sophistication. The right accent may conceal profound ignorance and ineptitude. In Britain, the substance of what you say is not quite as important as how you say it.

12

Low Connectivity
The deeper crisis of social disconnection

It's a gloomy Sunday. I'm stood on the edge of a beach gently rocking a pram containing my sleeping baby girl, looking on as my son builds a sandcastle with his mother. Children are playing. All is well in the world. About 20 feet to my left, climbers scale a rock face, aided by professional equipment. The air is cool but pleasant. This is that rare sort of day you often fantasise about when you're up to your ears in life's many responsibilities, working to deadlines, managing a household and being a parent from within an unforgiving fog of permanent sleep deprivation. When the cloud breaks and you locate yourself within the present moment, struck by that fleeting realisation you've been lost in thought, at the expense of now, you are filled with gratitude. But it's never long before you descend the rabbit hole once more.

We are not really at the beach, you see. Nor are we standing by a rock face. We are stood in a shopping mall which is doing its very best to simulate for us the experience of being outside. Nothing here is real. The sand is synthetic. The light is unnatural. The air is conditioned. On the concourse stand statues of fawns and small birds. In the car park, plastic flowers hang from artificial trees. The potted plants are fake. This environment has been carefully curated to create the illusion of nature and once glimpsed, it becomes hard to unsee its falseness. Even the police presence is cardboard.

So much of what constitutes human experience today is entirely simulated. We increasingly derive our sense of community from gadgets and contraptions which act as substitutes for genuine human contact. We give ourselves to digital applications which interact with us to simulate the intimacies found in human relationships. At night, we soothe ourselves and our children to sleep with the artificial sounds of nearby conversation or running water. We consume dietary supplements which replace the essential nutrients which come naturally from sunlight and energy products which promote alertness in the absence of adequate rest. Even our most intimate moments are increasingly enabled by technology. There is clearly a great desire to connect with nature, each other and ourselves but increasingly, this demand, which is central to our wellbeing, is supplied by synthetic produce or facilitated by technology.

This falseness also extends to human relationships that once occurred naturally as a matter of practicality in many communities. The adage that it takes a village to raise a child is rooted in the understanding that we function optimally in close-knit communities where we feel supported, integrated and useful. Today, society is awash with mentors, life coaches, counsellors, self-help gurus and influencers to whom we turn in moments of doubt, pain or crisis. Yet with more means to connect with others than ever before, why do so many still feel so alone? Why do so many people feel unsafe, insecure and unfulfilled? And why do we see an explosion in mental health problems at precisely a time when most of our basic material needs are being met. Might poor proximity to our fellows, in an age of seeming connectedness, be making us ill?

There is no substitute for human touch, a warm embrace, firm eye contact or a calm, reassuring voice. Each produce the essential hormonal cocktails which allow us to attach to others and our environments, become motivated and experience gratitude. It follows that the demand for synthetic and technological imitations arises because the authentic alternatives on which they are modelled are in shorter supply than at any time in our history. And in the absence of the essential ingredients necessary for the production of wellness, community cohesion, purpose and meaning, we reach out for whatever we can get our hands on to quiet the nagging feeling that we are in some way incomplete – oblivious to the tragic reality that the comforts to which we turn for a momentary reprieve are the cause of our deep unfulfilments. There is a way out of the simulacrum, and into a realer existence, but not everyone can afford the admission price. The less money you have, the less able you are to purchase your way out of the sensory obstacle course and the more likely your social connections are to become frayed as a result.

Social connection is everything in life. Many of the instincts, cognitive capacities and emotional responses which ensure our survival early in life are activated by the interactions we have with other people, emerging from attachments we form, and beliefs we develop, to and about the world around us. Think of the skin-to-skin contact between a mother and her baby. The sense of security elicited in the child by her loving embrace. Even before birth, we listen from the womb for signs that we are safe, secure and loved. As we grow, we develop additional attachments beyond our immediate caregivers, mimicking the speech and body language of our peers in nursery school, adopting new interests, attitudes and behaviours which cannot be accounted for by how we are parented. Our sense of independence develops, bringing us into conflicts, which, depending on how they are resolved, leads to the formation of strategies and values which subtly shape who we will become.

Some social connections are primary, like family and close friends, beginning from birth and sustaining us throughout our lives. Others are secondary, such as peers, work colleagues, teachers or schoolmates. Those individuals to whom we feel bonded in some way, through a shared endeavour, like college or a job, or interact with regularly, like a baker or a bus driver, become symbolic of the breadth and depth of our social connections. The more we have, the more we can nurture. The more our connections are nurtured, the more confident and resilient we become.

As well as promoting the development of empathy and a growing repertoire of dynamic social skills, relationships are educational and instructive. Every brush with another human being, no matter how seemingly inconsequential, can teach us something useful about ourselves, others around us and the environments in which we co-habit. Even if all we glean is that we don't enjoy being around a certain person, we still learned something. Whatever nugget of wisdom we gain, whether we realise it or not, is added to an accumulating bank of practical, actionable knowledge. A roughly drawn map to which we will endlessly refer as we navigate the imperfect, unpredictable terrain of our lives. Social networks and connections are central not only to our survival as individuals and as a species but also to the quality of our lives, the resilience of our families and the cohesion of our communities while we're here – and long after we are gone.

Our lives are improved immeasurably by merely interacting with one another. A trip to the gym is often time to unwind mentally but our physical performance and mental endurance are improved by training with others. We may nip out for a walk, hoping to declutter our minds, but find that a stranger's smile or kindly holding a door for someone places us on a surer mental and emotional footing. Someone referring to you by name or remembering something from a previous interaction can elicit a sense of emotional satisfaction which is palpable but also very difficult to describe. Words of encouragement help us work harder. The local barista remembering how you take your coffee evokes a comforting sense of familiarity. A local café owner throwing on an extra dollop of grub at no extra charge because you look tired may energise and inspire you. Every interaction lays the foundation of a natural rapport and trust from which a wider sense of safety, security and wellness can emerge.

To understand the malaise gripping post-industrial Britain, one must view the matter not simply as economic in nature but also through the lens of social connection. That is what deprived communities are often deprived of. You only have to contrast those communities regarded as deprived with those associated with affluence and prosperity to see there is more than simply a disparity in income.

I think of the sleepy little hamlets I visited while promoting my first book. The beautiful, tight-knit towns where the community's central focus was not food banks but on setting up an annual book festival – something no self-respecting town can do without, apparently. Villages like Birnam in Perthshire, famed for a tree so old it was allegedly written about by William Shakespeare. I recall the bright-red phone box on the corner, across the road from the Birnam's arts centre, a thing of beauty, wrapped from top to bottom in white, twinkling fairy lights. The hotel doors that were never locked. The front door upon which handwritten instructions on how to get in, in the absence of staff, was prominently attached. The single off licence and one or two pubs. The locals, many of whom were on first-name terms with each other. And I recall the dilemma of being unable to locate a bin to discard my cigarette butt, as due to the immaculate condition of the pavements, the thought of tossing it in the street filled me with panic and dread of judgement.

In stark contrast, the exterior of the Saracen Bar in Possil, where I chatted with locals while filming our series, was covered in cigarette ends, and the bin in which they should have been discarded had been torn violently from the wall. The local community centre was a relic from the 1980s, bearing more of a resemblance to a factory or a prison than a welcoming public space. Possil had no central preoccupation; the community was fragmented and aesthetically characterised by asymmetric disrepair and dereliction. Are people in Birnam just better than people in Possil or is there something deeper going on?

In post-industrial communities of the sort often discussed in terms of poverty and deprivation, social connections are under increasing strain and people have grown more distant from one another. An unbalanced housing market creates residential instability, forcing people to move home more often. The migration of secure jobs abroad has led to a precarious labour market where flexibility for employers spells endless uncertainty for workers. These conditions are not conducive to social connection. A confluence of economic uncertainty and social insecurity constrains the human ability to form and nurture the social bonds so vital to individual prosperity and community cohesion.

Then there are the coping strategies people turn to when they become socially disconnected. In many communities, the primary economic activity taking place is the purchase of goods and services, which ferment and enflame the social problems associated with them. Gamblers hope a big win may help them clear their debts or move to a better area. The old and infirm spend what's left of their time in pubs. Youths gather by off licences, hoping a weak-willed old-timer will buy the cargo that takes the edge of the boredom, the sense of being futureless and the ubiquitous threat of violence. Single parents fret at home, comfort eating through the shame, as the debts of servicing the skyward cost of living spiral up around them. In deprived communities, the bad habits that create temporary emotional relief are all some people have when their wellbeing is compromised. For others, as we have already explored, this disconnection becomes acute and constant, driving them into drug addictions and homelessness. But when the forensic reports are done, and the death certificates are published, no reference to social disconnection will be made.

Over time, the coping strategies to which we turn in our sense of disconnection often become self-insistent, rendering us further isolated in a problem. Parents withdraw from children under the duress of sleep deprivation. Children withdraw from school when they feel inadequate or threatened. Addicts withdraw from society completely, retreating to the margins.

In every case, life's adversities require support to traverse. Invariably, this support comes from other people, based on the social connections that are available to us. In communities where social problems are rife, it follows that social connection is all the more important. When elderly people reminisce about the good old days, they aren't being nostalgic about having more money – they had less. What they are lamenting is the sense of belonging that social connectedness brought to their lives. So connected did they feel that they were comfortable leaving their front doors unlocked or their children with neighbours.

Post-industrialisation is often framed in employment terms; traditional industries were wound down, creating mass unemployment and economic displacement which contributed to social problems, like crime and drug addiction. High-rise housing became synonymous with dereliction. Many developments were pitched to poor areas as progress but instead unlocked the back door for the unwieldy forces of gentrification. In the analysis of the effects of deindustrialisation, too little is made of the abrupt disruption and thoughtless dispersal of individuals and families who had once lived side-by-side. Many of these communities were ripped up by their cornerstones, with little understanding of the possible longer-term social consequences. The key aspect conspicuously absent from the mainstream post-industrial mythos, illustrated so often by the iconography of demolitions and mine and shipyard closures, is perhaps the most pivotal: the needlessly violent severing of vital social connections forged over decades that once bound individuals, families and communities in tight-knit, resilient groups. Connections laid down over generations that granted a sense of belonging and shared history in which people could orientate themselves and guide others.

There are serious health implications where poor social connection is concerned. Social connection improves physical health and mental and emotional wellbeing. It is, in many ways, a panacea for all sorts of problems, dilemmas and ailments. When we feel connected, we feel well; when we don't, we become prey to all manner of maladies.

Feel-good hormones like serotonin, oxytocin, endorphins and dopamine are central to emotional regulation and underpin the emotional states we associate with improved mental health. As with running, lifting weights or making love, human interaction plays a key role in our wellbeing. When we feel connected to a community, whether that be our family, friends, neighbours or just like-minded people, hormones are released, altering how we feel. Across many studies of mammals, from rats to humans, data suggests that we are shaped quite profoundly by our social environments and that when our social bonds are threatened or severed, we suffer. Data also shows that the brain makes little distinction between physical pain and social pain. That's why codeine is as a good at numbing toothache as it is at taking the edge off boredom or isolation. In fact, social connections are so fundamental to our wellbeing that isolation and disconnection are as bad for our health as smoking and drinking. Social connection strengthens the immune system, helps us to recover from illness faster and can even lengthen our lifespans. Feeling more connected produces lower levels of anxiety and fewer depressive episodes as well as improving self-esteem and promoting empathy and trust, and thus, greater social cooperation.

The widespread misconception that the individual is the bulwark of society is undermined by the science of social connection. We acquire language by listening to people speak. We learn to compromise by sharing resources with others. And we learn how to love by being loved. Where social connections are frayed or severed, individuals, families and communities suffer because the opportunities for more nurturing forms of behaviour to be modelled and adopted are reduced. This does not mean we do not connect or attach – it means we connect and attach to values, behaviours, attitudes and beliefs which are not always conducive to long-term health and wellbeing, and thus, they undermine the potential for greater social cohesion.

In the post-industrial period, as many communities vulnerable to changes in society's economic structure have become synonymous with poor educational attainment, unemployment, social immobility, ill-health and addiction, the demand for social connection is increasingly supplied by the public, private and third sector, with mixed results. Mentors simulate companionship for troubled young people. Counsellors offer fellowship to those with mental health problems. And teachers are increasingly asked to provide care, compassion and emotional support in an educational setting, which may be lacking at home. Meanwhile, social media (the commodification of our instinct to socially connect) has proliferated to the extent that we have become enmeshed in algorithms which are not designed with individual or community wellbeing in mind.

Increasingly, it appears these platforms work against our better natures and attack our vulnerabilities, incentivising unhealthy levels of conflict and self-concern which often precede slides into depression and anxiety – while keeping us logged in for the sport of advertisers. Has Facebook ever showed you a 'memory'? I don't know about you but I tend to cringe when I am forced to read something I said ten years ago, yet Facebook persists in revealing these memories to me regardless. The cynic in me reckons the ploy here is actually to induce a negative emotional state, one which may result in a sudden need for validation perhaps brought about by a long, self-regarding status or highly filtered, post-workout photograph. Likewise, Twitter is presented as a platform which connects us to the world, but no matter how I curate my feed, it tends to prioritise the outraged brain-droppings of people who live within a few miles of me. It constrains what I see and with whom I may connect.

Social media is sold to us as a means of connecting and of communicating, but incentivises us to think and behave in ways that leave us feeling isolated and misunderstood. The collective misconception around social media is that we can wield its profound power to change the world when, in truth, we are being overpowered by it as individuals and as cultures. It is only when we peek our heads out of Plato's Cave, momentarily glimpsing that we have confused truth and connection with a compelling shadow projected on the wall, that we experience a welcome, but fleeting, sense of relief and a glimpse of social media's unrestrained menace.

Studies regularly find that social media users struggle to regulate how often they use their socials. But many likely haven't given full enough consideration to what this really means – there is an involuntary aspect to how many of us end up online in the first place. At the heart of much of our activity, our sense of what is going on and our role in participating in that, lies a very subtle and telling delusion that we are acting autonomously when often the opposite is true – we are unconsciously scratching a psychological itch.

Like addicts, social media users often underestimate how much time they really spend 'using' and if ever challenged about it, can get defensive or dishonest. Much like other addictive behaviour, social media use affects attention, emotional processing and decision making. The immediate rewards our brains detect when using social media create self-insistent compulsions to continue using and we often persist despite the harm we know it causes us. What often eludes us is that, just as with drug addicts, alcoholics, compulsive gamblers or people who struggle with emotional eating, being deprived of social connection is what makes us more vulnerable to addictions of all kinds.

While social media is a relatively recent phenomenon, the commercialisation of human suffering, arising from inequality, is not. Communities which have been deprived of vital resources historically and more recently due to austerity are more susceptible to the pitfalls found in a poorly restrained free-market system. Their vulnerability is two-fold – as workers, they are increasingly disposable and as consumers, they are increasingly lucrative. They lack the means to bargain for better pay and conditions and they do not possess the means to temporarily purchase a way out of the consumerist labyrinth. Within capitalism, they become sitting ducks. The products and services which induce or enflame their mental health problems are sold as solutions to (rather than drivers of) their ills and often become the only means by which increasing numbers can simulate the sense of belonging and security they crave.

As well as the benefits of natural social connection for the individual, the wider community functions better if it is networked. Think of natural social connection in a community like Wi-Fi – the stronger the connections, the better the experience. In poor communities, connections are unreliable and intermittent. Nothing in a deprived community is long term, and this can be inferred from the fact they have historically been referred to as 'schemes' or 'projects'. The only continuity is uncertainty. Indeed, it is this incessant sense of low-level chaos which acts as an incentive for many to uproot and leave, thus creating an additional dimension of ceaseless upheaval as well as a constant migration of skills and attributes. They are constantly being stricken down, berated, condemned and harshly judged while having their roots pulled from beneath them. These are not conditions in which people can thrive.

Every time high streets fail, or a tower block is demolished, or a library closes, a section of an already vulnerable community is set back, displaced or dispersed. An area of local life over which locals once felt a sense of ownership, or in which they once took great pride, then falls into disrepair or is sold off to private interests. It's like trying to build a house of cards on soft sand. Over time, even the most active, informed and positive people grow cynical and demotivated. When new communities are designed, the science of social connection is never considered, despite the available evidence that it is essential for wellness and harmony.

Deprived communities are not inherently violent or dirty or run-down. These problems arise from the wounds they sustain. The impact is profound at the individual and social level as the community loses the ability to orientate and organise itself. Informal social controls break down. Without them, a vacuum opens into which more formal controls, like policing, or less savoury informal social controls, like criminality, insert themselves. The 'don't grass' mantra is a powerful form of informal social control, one that is rooted in a pathological distrust of authority. But many other forms exist. The expectation that one must respect one's elders comes with no formal punishment. but remains something that most people feel. Informal social controls arise naturally, as part of a community's culture.

In one community, the expectation is that rubbish must be binned, while in the other there is a sense that binning rubbish is pointless because the community is a mess. In one community, you wouldn't think twice about calling the police, while in the other just the sight of a cop car can plunge some into anxiety. Social connection and informal social controls are central to a community's capacity to cohere and prosper, to how the community looks and feels. But they are also pivotal to a community's ability to fight back. To rebuild. To redefine itself.

Perhaps the most striking example of informal social control in the UK can be found in Liverpool, where a mass boycott of the *Sun* newspaper – a profound demonstration of community strength in the face of the publication's lies following the Hillsborough disaster – has been operative for decades, despite having never been formalised. You still can't purchase a copy of the *Sun* – not even in a typical WH Smith at a train station – and asking for one anywhere else will raise eyebrows. There exists no law prohibiting people from buying or selling the *Sun*. But it is well understood that in the city of Liverpool, the newspaper is to be avoided. This is not simply an act of retribution but a response which provides a powerful example, deterring others from aggressing on that community.

Informal social controls are as essential to communities as food and shelter are to individuals. They provide a sense of what is expected of us and of what direction we are travelling. They may operate on various scales and across vast distances. As long as there is one issue, objective or narrative to which a community can attach, anything is possible.

But equal to this is the capacity for damaged communities to attach to harmful attitudes and behaviours. This is often attributed to individual choices but is in fact symptomatic of frayed or severed social connection at community level, arising from poor governance. Gang violence, obesity, co-morbidity, distrust of authority, fatalistic attitudes to health, education and social mobility, and the over-arching belief that nothing will ever change are not evidence of fecklessness but of social injury. But perhaps the most ominous development in recent years is the worrying rise in suicides – across all age groups.

The great leap in our understanding of suicide as a social problem rather than individual sin or moral failing came in the nineteenth century. French philosopher and founding father of sociology Émile Durkheim, in his landmark study *Le Suicide*, published in 1897, identified that socioeconomic conditions, and their impacts on human health, relationships and wellbeing, were driving a sudden spike in the French suicide rate as the country made its tumultuous transition from a traditional agricultural society to an industrial capitalist economy.

Durkheim was committed to sociology as an empirical endeavour and *Le Suicide* was revolutionary in that it was the first body of sociological research to use statistical data. He identified what he saw as four main types of suicide, each attributable to a specific alignment of social circumstances. Durkheim's system was based on his contention that a person's risk of death by suicide was relative to their level of social integration. That the quality of a person's social ties with family, friends and work associates was central. What these social bonds provided – and what those who attempted suicide often lacked – were opportunities to access a collective consciousness. A means to locate themselves within a broader sociocultural context from which they may derive a sense of purpose, meaning and a rough design for living that may sustain them. A pool of knowledge and experience to draw from and, indeed, contribute to. If an individual failed to integrate socially, or even if they became over-integrated (think of extremist groups), then the risk of suicide increased. While Durkheim's system is certainly indicative of its time, it remains instructive in that it correctly presents suicide as a complex multifaceted issue. Sadly, like so many social problems, suicide is still regarded as an individual choice and not a symptom of a deeper social sickness.

While death by suicide does not discriminate and impacts people from every background, there are certainly risk factors. In Britain, people who live in more deprived areas – where there is less access to services, work and quality education – are more at risk of suicide. According to analysis published by the ONS, those among the most deprived 10 per cent of society are more than twice as likely to die by suicide than the least deprived 10 per cent. While suicides have declined in the last 30 years, they rose to a 16-year high in 2019. People who work as carers and low-skilled workers have a significantly higher risk of suicide than those in other occupations.

Conversely, individuals working in roles as managers, directors and senior officials – the highest paid occupation group – had the lowest risk of suicide. Among corporate managers and directors, the risk of suicide was more than 70 per cent lower for both sexes.

The reasons why people think about or attempt suicide are exhaustive. A salvo of statistics, no matter how detailed, obscures the broader picture and makes it easy to lose sight of the largest determining factor: a person's social class. This remains the biggest predictor of risk. Given the suicide rate remains a key metric in determining the broader health of a society, it's skyward climb in recent years serves as a stark indication that something has gone very wrong. While those struggling with mental health crises are told that no matter what may be going on in their lives, suicide is not the answer, this well-meaning platitude is often issued from a loftier vantage point than the truly beaten and destitute will ever experience.

It is when you contrast the fragility of a negatively charged social network with the relative strength and resilience of one which is more positively charged that you begin to understand that the myriad advantages often attributed by the affluent to their own merit are in fact the socioeconomic manifestations of how well 'connected' they are.

13

He Who Pays the Piper Calls the Tune
Understanding why media is so bad at class

> 'We believe that concentration within news and information markets in particular has reached endemic levels in the UK and that we urgently need effective remedies. This kind of concentration creates conditions in which wealthy individuals and organisations can amass huge political and economic power and distort the media landscape to suit their interests and personal views.'
>
> *Media Reform Coalition, 2015*

On 19 February 2020, some days before the first lockdown was announced, Daniel John Hannan, Baron Hannan of Kingsclere, a writer, journalist and former politician, in a since deleted tweet claimed: 'Coronavirus isn't going to kill you. It really isn't.' On the same day, over on the pages of the *Telegraph*, leading business and economics commentator Jeremy Warner wrote: 'Not to put too fine a point on it, from an entirely disinterested economic perspective, the Covid-19 might even prove mildly beneficial in the long term by disproportionately culling elderly dependents.'

As Britain became engulfed by a pandemic our political leaders had had weeks to prepare for, talk of rising public debt, the cost of the rescue package and growing scepticism of the efficacy of lockdown, all underscored by a conception of the public purse as a household budget, rose with shocking Malthusian undertones.

Free-speech campaigner and *Spectator* columnist Toby Young then entered the fray on 31 March with an article in the *Critic* that rightly caused a vicious public backlash. Toby asserted that protecting the elderly from the virus was a waste of money. That it wasn't 'worth spending £185 billion to save them'. That we had to cut corners preserving human life to save the economy – like that theory had not already been road tested to disastrous effect during ten years of austerity. Young's view boiled down to this: if you happened to be old, and you happened to contract Covid-19, then tough luck because we couldn't afford to save everyone. He argued that those in their late seventies could only expect to live another few years – because average life expectancy in the UK is 81 – and that devoting so many resources to saving them did not represent value to the taxpayer. The problem with Toby's argument was that anyone who reaches the age of 79 can expect to live much longer; life expectancy is not fixed but slides on a scale the longer you are alive. It varies throughout your life because it is calculated as the number of years you are expected to live, given you have already reached a certain age. For example, a person born today in England is expected to live to 81 but a 79-year-old person today can expect to live until 91 – another 12 years. Young's argument, which failed to grasp this pivotal detail, read like the last thoughts of a man desperately attempting to rationalise the mass shooting he is about to commit. And with every passing hour, it seemed another middle-class commentator with no medical background to speak of, keen to frame the fast-moving events, emerged to suck a little more oxygen out of the room.

Good Morning Britain host Piers Morgan claimed in front of an audience of millions that 'we are all in the same boat'. I'm sure some of you remember that boat. That was the boat where you were paid £1,500 an hour for shouting over people in a television studio every weekday morning before going home to a lovely big property in London to bemoan the fact that coronavirus restrictions prevented you from swimming in your local private pool. Morgan's claim, that we were all in the same boat, was not simply contradicted by emerging evidence of vast disparities across the population with respect to infection and mortality rates, but also furnished those least at risk with a hard-luck story which revealed the chasm between them and the people trapped at home, forced to watch them on television.

Those who were living in poverty prior to the virus were by far the most vulnerable to it. A confluence of historic social, economic and cultural disparities across health, housing, transport, employment and income, as well as education and public trust in authority, drove up infections and deaths in poorer parts of the country, where crowded conditions made social distancing difficult and underlying health conditions associated with social deprivation complicated matters for patients and threadbare health services.

The class-based analysis on which Britain's emergence from the pandemic so clearly depended became increasingly drowned out by the overconfident bluster of media personalities and commentators, who were insulated from its worst effects. A concerted effort in sections of media to frame coronavirus as a matter of personal responsibility and resilience led to massive pressure building on poorer people to get back to work, which contradicted the public health advice they were already growing sceptical of. Working-class people were photographed and filmed constantly in parks, at shops or on public transport, often portrayed as acting unthoughtfully, when many simply had no choice. Even their permitted exercise time became news fodder, as Britain's chattering classes debated the wisdom of working-class people visiting parks during a pandemic. Was it safe? Were they being irresponsible? Should they be told to move on or even be detained on the grounds of public safety? Nobody was quite sure but one thing was certain – *Guardian* and *Spectator* columnists would get to the bottom of it all in between backyard chicken feeds and tending to their allotments.

Even as the disproportionate impact of Covid-19 on working-class people and the poor became undeniable, we seemed, as a nation, to be unable to grasp the reality that this was a crisis of social inequality. The pandemic, as it was often being discussed among media pundits and politicians, bore little resemblance to conditions in the poorest areas, where people were dropping like flies or falling completely off the map of health services. There was little sense of the daily nightmare experienced by disabled people, cancer survivors and those with serious mental health problems, who were left overexposed to the disease and the economic downturn which followed. Mainstream news increasingly framed events from the perspective of industry and government. There was no end to updates about the chaos going on behind the door of Number 10 – but often the reports centred on political rivalries and struggles for power and not the institutional intransigence of government departments.

We were living in a nightmare, where people at the top of key institutions appeared to have no idea what was going on – and even less idea of how truly out of their depth many of them were.

Then something strange happened. We seemed suddenly on the cusp of a momentous awakening. A lone BBC news presenter looked the country in the eye and spoke the truth. 'They tell us coronavirus is a great leveller – it's not.' Opening a highly editorialised, career-defining monologue, taking to task the misleading language surrounding the crisis and torpedoing the emerging narrative that lockdown was equally painful for everyone, *Newsnight*'s Emily Maitlis confirmed what millions had long since suspected: 'It's much, much harder if you're poor.' In doing so, she re-anchored the debate around those most affected, creating that rare televisual experience which causes you to leap up and out of your seat and haymaker the air.

Britain lay ripped right down the centre. On one side of the class ravine, women now trapped at home due to lockdown continued weighing up whether to flee their violent partners and risk being murdered or remain with them in the certain knowledge that more abuse was guaranteed, while on the other, mainstream commentators, working from home, lamented ponderously the good old days when 'politics wasn't a matter of life and death'. On one side of the ravine, people with drug problems continued to dance with death, more remote from health services and support groups than ever, using the same illicit substances that had only recently killed their family and friends, while on the other, hustle-culture advocates tweeted that if you didn't come out of lockdown with a new skill, side hustle or more knowledge, you 'didn't ever lack the time, you lacked the discipline.' In one world, low-paid workers turned to their under-resourced unions for support in fending off employers pressuring them to risk their lives (and the lives of others) by returning to work, while in the other, an upper-middle-class couple used the pages of a national newspaper to bemoan the high rate of council tax they had to pay on their

£1.5 million home and how the lockdown had placed their £120,000 kitchen renovation in jeopardy.

By that point, only an idiot could have denied that Britain's historic social inequality was not only the decisive factor in our haphazard response to the virus but also in much of the vacuous media discourse which arose from it. Britain, however, is in no short supply of idiots in high-ranking positions and so debates about whether hiring a cleaner was a feminist act ensued. Even the well-meaning, relatively harmless twats became torture after a few weeks, appearing increasingly twee and disconnected, referring to gardens, cars, nannies, cleaners, savings and every other trapping of middle-class life as if they were everyday things. If only the social ignorance of the well-to-do, exhibited so frequently on television and in print, had been in shorter supply. Sadly, unlike basic life-saving equipment, the privilege was all too abundant.

How many of you, with no direct experience of poverty, have been surprised, shocked or even slightly repulsed by some of what you have read so far in this book? How many of you, lucky enough to have never had to turn to the Welfare State, would have thought, for example, that a public body like the Department of Work and Pensions was in the business of re-traumatising and placing at further risk vulnerable women fleeing male abuse? Would it have ever occurred to you that local authorities can kick the vulnerable and elderly out of their homes and onto the streets for the crime of not remaining in them for a few weeks? Or that people seeking help for drug problems have been forced to choose between residential treatment or homelessness? If you are among those who are surprised to discover that cruelty is perpetrated to such an extent by public services in a wealthy, modern, civilised Western democracy, then perhaps it's time to ask yourself why.

There is no shortage of media coverage where the poor are concerned. Every other day, news is awash with stories of food banks and child poverty statistics. Soap operas dramatise the feckless lives of working-class people every weekday evening. Controversial reality shows like *Benefits Street* have, through the years, attempted to bring you all a little closer to the action. Yet somehow, despite more coverage, commentary and content than ever before, many who would regard themselves as socially conscious, knowledgeable and informed still fail to grasp the symbiotic relationship between their own prosperity and the unnecessary hardship of others.

Sure, they may be able to rattle off some top-line statistics about the housing crisis. They may express genuine dismay at the widening educational attainment gap. But do they understand that a chronic lack of access to affordable homes for millions of people in Britain adds more value to their own properties? Do they truly grasp that their middle-class children are far likelier to succeed in education and the labour market than poorer kids in part because kids from poorer backgrounds are being held back?

It is in this ignorance of the structural nature of inequality that politicians are given a free pass as poverty is broken into bitesize subgenres, each with its own poverty industry growing up around it, while the real story of its systemic nature is rarely told. And, as we have explored with respect to issues like homelessness and the drug crisis, it is this that allows pundits without a clue to confidently assert lazy stereotypes, falsehoods and blatant lies purporting to account for why the lower classes do less well, which then take stubborn cultural root.

The inequalities outlined so far in this book are not simply overlooked by a media dominated by certain social classes – they can only exist to the extent they do in Britain today if their true nature is actively obscured. And yet it is only when injustices are exposed by media that narrow perceptions begin to shift, and that overdue progress may occur. Think of the MPs expenses scandal of 2009, or more recently the second jobs furore in 2021. In both cases, these took place in a culture of entitlement enabled by soft-touch rules and lack of regulation of professional politicians. For decades, this culture was widely known to exist and was even accepted, often with great resignation, as normal; it wasn't as if your MPs entitlement to hundreds of thousands of pounds in expenses every year, or to a second, third or fourth job, was some great secret. What endowed both scandals with their respective revelatory qualities was not the culture of entitlement itself but how this was re-contextualised. The banking crisis and the pandemic, respectively, cast the parallel universe many with power and influence had been living in, as millions outside that privileged bubble suffered, in a new and damning light. Seemingly old news was suddenly rendered shocking as a direct result of how it was presented and framed by the press. The depth of discussion it provoked. The intensity of scrutiny applied. Only then did the wider public begin paying attention. And only then did it become possible to hold the privileged to account and create the demand and momentum for change – that is the power of a free press.

Emily Maitlis's monologue was so affecting because such an occurrence, where a famous journalist knowingly places themselves in harm's way to defend ordinary people being spoken over by the privileged, is so desperately rare. And even then, she paid a price, as sections of the 'free press' attempted to frame her evidence-based argument as proof of media bias. Much of the backlash against her was generated by the same media forces which have waged open war on vulnerable groups while publishing puff-pieces every other day, imploring us to fawn over the royal family, celebrities and tax-evading captains of industry.

To ensure fairness, equality and accountability, and that people of all social classes get the fairest possible hearing, many of us look primarily to news and media institutions, which we entrust to make sense of it all. But even the publicly owned broadcasters, like the BBC and Channel 4, which today exist to provide a counter-balance in the heavily deregulated, commercial media environment, are, themselves, mired in the cultural effects of social inequality.

Numerous surveys on the class backgrounds of media professionals show they are overwhelmingly upper-middle class – meaning they have more in common culturally and economically with the wealthy than they do with the poor. Research conducted by Sam Friedman of the London School of Economics in 2019 found that 67 per cent of Channel 4 staff – regarded as a more representative broadcaster due to greater gender and ethnic diversity – had parents who worked at professional or managerial level and only 9 per cent identified themselves as coming from a working-class background. The BBC was moderately better, which is to say still atrociously middle class, with 61 per cent of staff reporting they were from upper-middle-class backgrounds. Seventeen per cent of staff and 25 per cent of BBC management went to private school – well above the 7 per cent average for the population as a whole.

Working-class people in media organisations can usually be found serving food in the cafeterias, working on the doors or watching over the car parks. When you enter the building, they'll be sweeping the floors or cleaning the toilets, usually in uniform. The middle classes occupy the lower and upper tiers in the mid-section of the structure, with greater levels of autonomy and authority, and are permitted to dress however they please, while acting as the public face of the institution. Meanwhile, the people at the very top, who set the institution's agenda and decide which middle-class people to employ to carry that agenda forward, are rarely seen. British media institutions are, in many ways, a tribute to the British class system in its most traditional, linear form. What hope is there that we may, as a culture, arrive at a deeper understanding of the true nature of Britain's structural imbalance when our free press and media organisations are dominated by the middle and upper classes, whose economic interests are closely aligned and whose assumptions, prejudices and even bigotries are often betrayed by their workplace practices and multi-media creations.

An optimist may point to instances where the press has acted in the public interest, such as the expenses scandal of 2009 or Emily Maitlis's monologue, and proclaim with misty eyes that these stand as evidence of the robustness of Britain's free press. They may bring up the *Daily Mirror*'s opposition to the Iraq War or the wonderful *Private Eye* as proof of media plurality. But a cynic would argue, correctly, that every example you might cite of a publication or broadcaster going against the general grain is an exception to the rule and that while aspects of it remain 'free' in theory, British media is less 'free' than it has ever been because a small number of extremely wealthy people have been allowed to acquire so much of it.

The Media Reform Coalition, in March 2021, published a report which showed that just three companies (News UK, Daily Mail Group and Reach) dominate 90 per cent of the national printed newspaper market (up from 71 per cent in 2015). When online readers are included, these three companies dominate 80 per cent of the market. In the area of local news, just six companies (Gannett, JPI Media,

Reach, Tindle, Archant and Iliffe) account for nearly 84 per cent of all titles. Two companies, Bauer and Global, now control nearly 70 per cent of all local commercial analogue radio stations and 60 per cent of national commercial digital stations. In many ways, this concentrated ownership and the disproportionate influence it grants some is analogous to landownership, except in this instance, it is not the ground we walk on but almost everything we see, hear and read that is controlled by a small, select group.

While the deregulation of media has resulted in a massive expansion of choice, power is concentrated in the hands of a few. The media infrastructure in the UK acts increasingly like a soda dispenser you might find at a cinema complex – it serves 50 different types of pop, which all taste a bit different, but they are all comprised of the same basic ingredients and squirted into your cup through the same tap. When we consider that commercial news and entertainment operate from the same underlying principle – supply, demand, profit – this analogy no longer seems too far-fetched.

Incidentally, a viable media outlet, despite its gradual decline in the face of the internet, remains a must-have accessory for many who count themselves among the super-rich. Some even run their most famous products at a loss, meaning the incentive to remain in the market may not necessarily be financial. Given these wealthy individuals don't become powerful by spending their money frivolously, we must then ask: what do they expect in return for their investment?

Well, they expect to live, as a rule, rent free, in the heads of the politicians who take the big decisions that affect our lives. They also expect a compliant media environment where a fixed range of topics is the subject of lengthy, often exhaustive, discussion, creating the impression of plurality but where the seemingly lively debate is always constrained within certain ideological parameters. If you own a few big daily newspapers on every continent and a couple of giant news networks, and so have access to hundreds of millions of readers and viewers (voters) 24 hours a day, then a politician, hoping to get elected, or to stay in power after a disastrous tenure, might decide, for example, to tinker with some parts of the legislation that regulates or restrains your capacity to expand your power. Hell, they might even invite you in to write some of the legislation yourself and maybe even tell you what kind of country they'd like to see emerge after the next election. Who knows?

Today, when you hear some people talk about a free press, they don't mean simply that publications and broadcasters should be free to hold the powerful to account. Sometimes, what they also mean is that true liberty is also expressed in the right of the rich and powerful to own as much of the press as they desire – and that any attempt to constrain their ambition is tantamount to tyranny. This is a complete distortion of the notion of a free press.

What this all boils down to is a media which, through ignorance, deliberate omission or motivated reasoning of the facts, is generally incapable of assessing the fundamental scandal of British society – class inequality – because such a scandal would have far more serious implications for wealthy media proprietors than they are historically accustomed to. Worse still, anyone in the public eye who intends to speak honestly about this, or even with temerity to bring it up, can expect a public lashing from powerful sections of the 'free press' – most of it bought and paid for by wealthy men who'd rather you didn't worry your pretty little head about such lofty topics as concentrated wealth and power, and instead, directed your anger at single mothers, immigrants and refugees.

What is deemed newsworthy, entertaining or informative is generally assumed to be a matter of the middle-class sensibilities of audiences and media professionals. But often, the ideological agenda underpinning a paper or broadcaster's output is driven by interests well above their paygrades. The most senior middle-class television presenters, producers and commissioners remain, in some way, beholden to someone else further up the structure. Even publicly funded outlets like the BBC and Channel 4 must contort themselves in order to compete with the private sector and so much that needs to be discussed with respect to class inequality is placed out of bounds. Meanwhile, newspapers and broadcasters are bought up, merged, asset-stripped, downsized and forced to depend increasingly on the revenue generated by advertising – an industry where lying is professional virtue. In this environment, even the most virtuous journalists, presenters, producers and directors stand little chance.

The absence of consistent, running coverage of the interconnected, systemic nature of class inequality in Britain is no accident. Its conspicuous absence is clearly in someone's interest. Those who most aggressively dismiss this notion, or who push back most vociferously against it, tend to be those in the deepest denial of how truly enmeshed in the injustice they have become. This media silence runs much deeper than forgivable ignorance arising from low proximity – it borders on complicity. Mid-level, middle-class media professionals look to their superiors at executive level like working-class kids do their school teachers: with unjustified reverence and a fear of consequences for stepping out of line. That is how classes are regimented in a system which is designed from the top down.

I work in media. I write books. I have a tabloid column. I write and present 'hard-hitting' documentaries. But I remain an outsider by choice because I understand the temptations. Rarely are words to any effect ever spoken which indicate what is expected of you in these industries but, rather like in the communities we examined earlier, an informal social pressure is palpable, growing more acute the higher your rise, which subtly directs your thoughts and behaviour in relation to certain issues, which must always be discussed in certain ways.

Where the lowborn are excluded from working in the media by virtue of the myriad inequalities they face, value must be extracted from them in other ways. Put simply, when working-class people cannot be formally employed to *create* news and entertainment, instead, they become it.

When news that *The Jeremy Kyle Show* was to be pulled abruptly off air in May 2019, the producer ITV could be sure a media storm would imminently make landfall. In the age of the algorithm, where speculation and gossip make headlines irrespective of their veracity, it is highly unusual, and extremely risky, for a massive media conglomerate to perform such a convulsive, and costly, public relations manoeuvre. Something was wrong and they knew it.

The following morning, it emerged that Steven Dymond, a 63-year-old digger driver, had been found dead by his landlady in the bedsit she rented him, ten days after taking part in the filming of an episode. An episode which, had it been broadcast, would have seemed no less ordinary to even the casual viewer. One which would have followed the familiar format in which participants were brought on under the watchful eye of host Jeremy, who proceeded to tease out their problems before offering his characteristically personal take. Often Kyle could be kind and sympathetic, appearing to show compassion for his guests. His occasional warmth formed a large part of the show's hypnotic appeal. Those experiencing difficulties over which they had no control, such as a bereavement, abandonment or infidelity, played the role of victims to Kyle's rescuer.

The Jeremy Kyle Show exhibited the occasional flourish of brilliance, creating many genuinely heartfelt and candid moments that spoke to human frailty and misfortune. But that wasn't the big draw for audiences. Any generosity on display functioned merely as a palette-cleansing prelude to the show's main ingredient – Kyle's caustic contempt for the undeserving poor. His legion of loyal fans – mainly single mothers, housewives and the long-term unemployed – remained glued to their armchairs in keen anticipation of their protagonist's Vesuvian eruptions. For it was when Kyle fully embodied the role of the abrasive inquisitor, in wild-eyed and abusive tirades, that this deathly, middle-class media vehicle truly came to life.

The Jeremy Kyle Show was a ruthless and predatory enterprise that banked on the precarity of its viewers and participants. It often scouted for contributors outside job centres. The production team also used Facebook to specifically target people struggling with personal issues, like family breakdown, relationship infidelity or even domestic abuse. And then, once in the studio, in full glare of the cameras, Kyle took them to task.

'I'm ill, I've got a back condition and I can't ...' said one woman backstage, before being cut off by Jeremy. 'I don't mean to be rude, love, but if you've got a back condition, why do you spend so much time on it?' he quipped, delighting in his misogyny. 'Come on, Charles, why won't you come on stage, then?' he sneered at another guest. 'Because I'm shy, I don't like cameras,' said Charles, visibly shaking. 'Well, that's great,' Kyle roared, 'because you're talking into one and your face is being broadcast to millions of people, so why don't you save me time and get out here and face the music.'

The media debate about the questionable ethics of *The Jeremy Kyle Show* is as old as the daytime mainstay itself. Perhaps the most prescient journalistic artefact produced concerning the show was *Guardian* journalist Carole Cadwalladr's report from an episode in 2008, which she attended undercover as a member of the audience. In the subsequent article, 'Behind the Scenes at Jeremy Kyle: When reality bites, it leaves deep scars', Cadwalladr discovered that one of the participants – an 18-year-old named Jamie, who took part in a DNA test to establish whether his partner's newborn child was his – suffered from paranoid schizophrenia and bipolar disorder but that researchers showed little interest when Jamie's step-mother, Karen, tried to alert them.

'It was so, so very wrong what they did,' Karen told Cadwalladr, 'It was almost like ritual abuse. And I wasn't allowed to see him beforehand! They kept him from me . . . Afterwards, when I saw him, when I hugged him, and he was crying his eyes out, he was absolutely shattered by the experience, and I was very fearful about what he might do to himself. I turned to a producer and I said, "If anything happens to my son tonight, you will have blood on your hands."'

'It is so wrong,' Cadwalladr wrote, 'I cannot believe how wrong it is. I am watching a vulnerable young man being publicly humiliated on a makeshift stage in a pub car park in front of his friends and neighbours and, for those lucky enough to receive the DVD for Christmas, the nation at home.'

Carole also noted one of the many remarks Kyle made about the vulnerable teenager, to the delight of the audience: 'You've got about as much get up and go as that pair of steps, mate, what is wrong with you? Open your gob and say something. Because I don't know, and I say this every morning, but I genuinely don't know the results. But right now, looking at you, you don't look the most trustworthy, if I'm completely honest.'

It subsequently transpired Jamie was telling the truth. But the seeds of doubt were by then sown. An ITV spokesperson later said that Jamie had been assessed by a qualified mental health nurse and psychotherapist prior to the recording, and that the show's director of aftercare spent a long time with Jamie after filming, with both finding no evidence of mental illness or risk of self-harming. But the damage was done. The fleeting gratification of Kyle's audience, hooked as it was on the tickly bit of anticipation prior to the scandalous revelations that punctuated every 60-minute serving of the show, had been satisfied. Its viewers, as well as the media institutions and advertisers that fed so gluttonously at the trough of this horrific exploit, had by then become perversely adjusted to the daily onslaughts, developing an ever-higher threshold for shock and disgust, and an insatiable appetite for still more extreme and rousing television experiences.

As news of Steven Dymond's suicide became public in the spring of 2019, the tragedy provided Jeremy Kyle with the ultimate reality-TV send-off. A tragic and deadly orgasmic climax. Dymond became the ultimate snuff tribute, sacrificed on the altar of one of television's cheapest, nastiest and most lucrative genres – poverty porn.

Many naturally attribute poverty porn's tackiness to the poor taste of its predominantly lower-class audiences whose lurid demands are simply being supplied. It was riveting for the same reason the scene of a car-crash is hard to look away from – people are nosey and the hope of witnessing something truly shocking is sometimes too hard to resist. But public fascination with poverty is not always a reflection of what audiences want but a measure of what is convenient and profitable to middle-class media professionals in news and entertainment, working at the behest of wealthy executives and proprietors. Viewing the poor and vulnerable through a distant, scrutinous lens for entertainment, or a misguided sense of superiority, has taken many forms throughout Britain's history. All that has changed are the canvases on which these images are projected and how rapidly they are disseminated culturally, shaping perspectives and opinions as they permeate.

The only reason a figure as absurdly uninformed as Jeremy Kyle could ever have been taken seriously as a broadcaster in the first place is because the media environment in which he emerged was, and remains, largely defined by its utter contempt towards the poorest and most vulnerable groups who do not possess the means to fight back. The death of Steven Dymond created a media feeding frenzy, as many of the tabloids, talk shows and media personalities which had previously sniggered along as accomplices while Kyle ploughed the depths of televisual depravity, recast themselves, somewhat conveniently, as concerned bystanders, strangely keen to hold him to account – and throw the scent of blood off their own hands.

In the days that followed, it emerged that Dymond's appearance was motivated by a desire to convince his fiancée, Jane Callaghan, that he hadn't, as she suspected, been unfaithful to her. Like almost all participants, Dymond was a fan of the show, and of Jeremy. He believed he and his partner might find a resolution to their relationship problems and remain a couple. But things took an ominous turn when it was revealed that Steven, who had maintained his innocence up until and after the result was revealed, failed the show's infamous 'lie-detector test' – a gimmick designed to create dramatic tension and sustain audience numbers throughout multiple advertisement breaks.

Since 2005, paper tough-guy Kyle had been a household name for fronting his Asda Smart Price version of the rate-winning US car-crash, *The Jerry Springer Show*. Springer rose to prominence in the nineties, his daytime bear-baiting becoming a globally syndicated phenomenon as the rust-belt trailer-park's answer to Oprah Winfrey. For over a decade, Springer repurposed lower-class, troubled and vulnerable Americans as workhorses to enrich himself and countless media executives, dragging television into a surreal and scandalous moral gutter. Participants were lured into television studios with promises of star treatment, swanky hotels and that coveted 15 minutes of fame while Jerry played the multifaceted role of concerned doctor, trustworthy counsellor, skilful mediator and reassuring friend. Viewers applied in their thousands, desperately seeking resolutions to their interpersonal adversities, believing with all their hearts that Jerry wanted to help them – oblivious that they were being viewed not as human beings but as a commodity.

Profit was pursued so ruthlessly and with such efficiency that not even the murder of a woman, battered to death by her former partner after the episode in which they featured was broadcast in 2000, was enough to get it cancelled. The only thing that killed Jerry in the end was low ratings.

Who in their right mind would ever wish to replicate such a thing on this side of the Atlantic? Enter Jeremy Kyle – supremely dim, auto-cued speaker of hard truths. As he strutted the stage every weekday morning, for 14 agonisingly long years, his personal, copyrighted brand of social justice, dispensed with vengeful, callous abandon, would have seen any GP, teacher or social worker struck off. What cheapened his antics further, if such a thing were possible, is that it was all an act. Kyle played a character, not unlike a professional wrestler does, turning from hero to villain and back again within minutes. His tone-deaf routine, rooted in unearned moral authority, was a carefully calibrated and cynical performance. His freakishly reliable talent for perfectly attributing blame before teasing the post-commercial-break cliffhanger, was a licence to print money. Kyle wasn't tough. Nor was he edgy. Indeed, public school-educated Kyle was rarely seen on set within three feet of a working-class person without a security guard present. Anyone with a degree of social experience could see through the straight-shooting, up-in-your-face routine – Jeremy Kyle is the guy who claps when the plane lands.

Endeavours of this low-brow nature are commonplace in television schedules and though they fall within the genre of 'reality TV', they have more in common with game shows, which are television's answer to the sausage – once you know how they are made, they become harder to enjoy. In television, it is widely understood that much of what ends up on screen is thoughtless garbage. The rationale within the industry is that the cheaper, nastier programmes generate the profit necessary to please shareholders and re-invest in more high-end, thoughtful and socially responsible programming (for middle-class people), such as award-winning primetime dramas depicting middle-class people having terrible sex in marbled wet-rooms before trying to kill each other. With its familiar formula, close-knit production team and logistically simple in-studio format, the running costs on shows like *Jeremy Kyle* are kept to a minimum. Like game shows, reality TV is attractive to executives because it generates a lot of revenue while requiring very little investment.

The discussion around the show's ethics, with respect to its premise, has always been something of a diversion from the exploitative economic imperative that underwrote it and all shows like it. Even Cadwalladr's prescient article failed to identify this. All moral and creative justifications for Kyle's bear-baiting, convincing as they may seem, are self-serving, post-hoc rationalisations: the show existed because it lined the pockets of television executives and provided countless opportunities for aspiring middle-class media professionals. In the year before its cancellation, *The Jeremy Kyle Show* is estimated to have raked in around £80 million for ITV. This accounted for around 5 per cent of their annual ad revenue. To put this in perspective, that's around £328,000 in advertising revenue every day, £6.9 million a month and £83.55 million if it were to run over the course of 12 months.

The famous lie-detector test was the centrepiece, around which much of the show's contrived sense of drama revolved. Despite being no more reliable than a throw of a dice, the lie detector generated countless opportunities in media – for spin-off news stories, YouTube clips and talking points – creating still more views, engagement and profit. Tom McLennan, an executive producer on the show, defended the tests when quizzed by MPs, claiming participants gave their informed consent prior to filming and that they were told the tests were not 100 per cent accurate, repeatedly. But he was also forced to concede before the Digital, Culture, Media and Sport select committee that, 'If you ask different people, you get different answers.' He added: 'I'm not a lie-detector expert.' Funny that.

Kyle's right-hand man, psychotherapist and director of 'after-care', Graham Stainer, whose primary role was to endow the show with medical legitimacy, later revealed to MPs that, 'Some people will fail this test yet they will be telling the truth.' He added: 'I totally accept that I don't know the percentage of success or the percentage of failure.' It's surprising how little anyone on the show really knew about the lie detector. It had become no more than an afterthought. And while the show's producers maintained that participants were made aware of the potential for inaccuracy, this warning played more like a knowing wink to the audience, for the fact remains that the show quite simply would not have worked in quite the same fashion without this cynical gimmick at its heart.

The premise of the show was holding people accountable. Wherever adversity or turmoil occurred in the personal lives of guests, blame had to be apportioned. The truth was always black and white. Guests were encouraged to antagonise each other in the hope it would heighten the drama. Heads regularly rolled, in the glare of camera lenses, for the sport of the crows in the crowd, bringing resolution to viewers at home.

But this seeming commitment to the principle of personal responsibility and the virtue of holding your hand up and accepting you are at fault seemed to fade ever so quickly when, after Steven's death was made public, tough questions were no longer being directed by Jeremy at the show's participants but by the press at the show itself. It was then that the truth suddenly became less black and white and more a nebulous grey area. Oddly, despite shouting in the faces of hundreds of guests for well over a decade about why they ought to grow a pair, stop making excuses and face the music, in the aftermath, as rumours swirled that he castigated Dymond on set, Kyle himself fell strangely quiet. The behavioural standards to which he held his guests did not apply to him.

Conveniently, Jeremy and his colleagues enjoyed the protection of ITV – a corporation where individuals are insulated from serious personal consequences by opaque legal personalities which limit their liability. Lucky them.

In November 2020, a coroner, in a pre-inquest review into the death of Steven Dymond, speculated that Jeremy Kyle 'may have caused or contributed' to the 63-year-old's suicide. Hampshire coroner Jason Pegg made Jeremy Kyle an 'interested person' for the inquest. 'It might seem ludicrous not to have Mr Kyle to give evidence to give his take on the situation,' Mr Pegg said. Kyle and ITV's legal team argued that Mr Dymond's 'upsetting experience' on the show was 'established fact'. The pre-inquest review heard how Dymond was 'booed and jeered' by the audience when the lie-detector result was announced, before being 'called a failure by the presenter'. Counsel for Mr Dymond's family, Caoilfhionn Gallagher QC, at the review, said Kyle was 'in his face' and even when he was 'at the point of collapsing' he was 'still being heckled'.

Ms Gallagher said his state of mind was known by the crew on the show, with a message sent on a WhatsApp group stating: 'Just so you know, he's still crying, he has just said he wishes he was dead.' The review heard how Mr Dymond was originally turned down as a guest on the show but was later accepted after being given a letter from his doctor. Dymond had been receiving mental health care from Southern Health NHS Foundation Trust. In his ruling, Pegg noted that he had watched *The Jeremy Kyle Show* episode in which Mr Dymond appeared. 'In that footage, it is apparent that Jeremy Kyle was aware that the deceased had previously been unable to appear on *The Jeremy Kyle Show* having been diagnosed with depression, for which the deceased had been prescribed anti-depressant medication,' he said. 'After the lie-detector results the deceased looked visibly upset. Jeremy Kyle adopted an approach where he called the deceased a "serial liar"; that he "would not trust him with a chocolate button" and made a comment, "Has anyone got a shovel?"'

That's what a free press looks like when reactionary billionaires set the news agenda every day. *The Jeremy Kyle Show* could only exist where there is widespread ignorance, prejudice and a lack of empathy, cultivated by powerful media institutions. Indeed, Jeremy Kyle's buffoonery would not have looked out of place on the pages of many of Britain's daily news titles, filled as they have been in recent years with spiteful, misinformed invective, designed to illicit hate and revulsion of the poor in readers. Vulnerable people with complex personal problems rooted in structural inequalities, who cannot sue for defamation because they are dirt-poor become the target of this misplaced fury when what they require is understanding and compassion – and what some of the owners of these papers need is long overdue dressing-down.

The Jeremy Kyle Show was merely the daytime TV branch of a British media landscape bought and paid for by elites who regard Britain as a minor piece on a free-market chessboard. A landscape in which powerful, wealthy individuals are enabled by governments to conflate the principles of free expression and a free press, with unfettered freedom to influence public opinion in accordance with their own narrow economic interests. Interests which can never be reconciled with working people.

The Jeremy Kyle Show was a lightning rod for the vulnerable. The very nature of Dymond's demise only lends credence to the fact the show attracted mentally unstable people. Indeed, the producers sought such individuals out as their frailties created the air of unpredictability, which made the show exciting. Kyle has since re-emerged, playing the victim card, claiming he was unfairly 'cancelled'. We wish, mate! This, from a man who has never managed to shut up for long enough to legitimately claim he's been silenced.

The ensuing debate that followed Dymond's death, dominated as it was by middle-class media professionals whose objectivity was hardly reliable, focused, as it had done for the previous decade, on the demonisation of the poor for entertainment, neglecting to zero in on the real story: countless, affluent, socially mobile, aspiring media and advertising careerists who profited from the daily abuse, coercion and exploitation of poor, mentally infirm and psychologically vulnerable people. From the low-level staff to the executives at ITV. The countless media publications that regularly ran coverage of Kyle's exploits in print and online. The many advertisers, targeting people on low-incomes with ads for everything from alcohol, rent-to-buy products, junk food and gambling. *The Jeremy Kyle Show*, in this context, represented the tip of a very large iceberg. A cottage industry of middle-class people, oblivious to the harms they are perpetrating, milking a human gravy train for everything it was worth – until a body dropped.

I look forward to the award-winning primetime drama about that.

14

The Goldilocks Zone
Meritocracy and the hubris of elites

Success is where preparation and opportunity collide, but while we can all prepare for an opportunity, one is never assured. And the quantity and quality of opportunities vary wildly, depending on where an individual happens to be situated. People who achieve success often develop the belief that their professional ascent was a simple matter of hard work and a positive mental attitude because that is what their experience tells them. We live in a culture which rightly celebrates success but which is less interested in those stories where years of preparation and graft lead to little but painful realisations that the stars may never align.

'Rags-to-riches' tales of people overcoming the odds are, sadly, exceptions to the rule but quarterly unemployment figures, annual drug deaths, homelessness statistics and the rising suicide rate are not as 'sexy' as the cultural construction that 'you can do anything you set your mind to'. Tracing back through their stories, 'successful' people partake in egregious revisionism, constructing the narratives which society demands and selects for. These narratives are then popularised, diffused through institutions and recycled as well-meaning platitudes and maxims which the aspirational use as templates in their own self-authored hero-journeys.

Nothing can exist without an environment to support it. This is as true of stars and planets as it is of plants and animals. Sheer celestial chance placed us just the right distance from the Sun, in what is often referred to as the Goldilocks Zone. The Earth's proximity from a flaming ball of gas is what makes our existence, our lives and perceived achievements and failures possible – yet we're supposed to believe Boris Johnson is the prime minister because he is the best person for the job. Have you ever heard anything so ridiculous in your life? Boris Johnson, a man with all the grace of a potty-training toddler. Basically, what you would get if thrush was a person. And yet, despite his relatively privileged life, Johnson in his bones believes he rose to the very apex of British society because he earned it. Because he deserved to.

So too do the countless other monied, landed, privately educated people who have risen the ranks to the great offices of state. What they all share is the belief, to varying degrees, that they ascended to positions of national consequence because they earned it. Of Britain's 55 prime ministers, 28 went to Oxford and 14 went to Cambridge, yet successive prime ministers have believed wholeheartedly that they were worthy, irrespective of the evidence that their ascents were determined as much by the favourable circumstances of their birth, and the choices laid out before them as a result, as by any inherent capacity they possessed. They are all, to one degree or another, purveyors of the greatest cultural myth of all: meritocracy. This powerful delusion that in the UK, an individual rises according to their talents, strength of character and work ethic, is the latest in a long line of privileged misapprehensions entertained by many in the country's upper echelons. Meritocracy is a wonderful ideal but when individuals who've never tasted desperation talk about it like it's real, it knocks society off its axis, as millions following that example draw inappropriate conclusions as to the true nature of their perceived successes and failures.

The myth of the meritocracy is fairly recent but the wacky, self-serving ideas of elites, and their institutional shoe-shiners, are as old as civilisation itself. For millennia, the assumptions of the wealthy and powerful have been road-tested on unsuspecting populations, often to disastrous effect. Much like Jeremy Kyle with respect to his hapless guests, their terrible questions, coupled with low proximity to social reality as it is experienced by the lower orders, compounded by their lack of accountability, have produced, and re-produced, endless toxic social and economic problems and pernicious, pervasive cultural myths and beliefs.

In 1798, *An Essay on the Principle of Population* by English political economist Thomas Robert Malthus predicted a nightmarish future where the exponential growth of humanity would outpace the Earth's capacity to provide food. 'The power of population is so superior to the power of the earth,' Malthus wrote, 'to produce subsistence for man, that premature death must in some shape or other visit the human race.'

Malthus believed that helping the poor through government intervention incentivised them to reproduce, thus placing a greater burden on society by exacerbating poverty. His influence led to the repeal of pioneering legislation dating back to the Elizabethan era which guaranteed food to the poor. Malthusianism became the basis of many myths and social stigmas around poverty and the poor which persist to this day.

While Malthus has, in some ways, become the poster-child for the bad ideas of elites, he was far from alone in his naivety. Indeed, many of Britain's celebrated intellectuals and beacons of enlightenment held and advanced similarly misguided ideas. John Stuart Mill, desiring to guard against the will of the great unwashed, proposed a democratic system in which citizens would be valued according to their level of education. 'Plural voting' followed from his conviction that the well-educated should be allowed to cast more votes, thus guarding against the tyranny of the majority. Calvinists believed those who rose to positions of national consequence were chosen by God, and Antisuffragists that women would be 'masculinised' by participating in political life.

Imperialism, racism, sexism and, later, renaming debt 'credit', the idea that predatory homosexuals might prey on 17-year-olds and so their age of consent should be higher, tuition fees, right-to-buy and trickle-down economics are all ideas either dreamed up or imported from overseas by highly educated, sophisticated and wealthy individuals. Those who were aided by centuries of unbroken social connections and networks, nested within pockets of immense wealth and privilege, and who believed they got to the top because they were worthy.

Some of these mad beliefs are more forgivable than others. And while it is always unwise to judge old ideas by the ethical standards of today (I suspect and indeed hope much of this book will age just as ungracefully), many of these now discredited ideas were actually resisted at the time. Many of the notions held to be true by current members of Britain's establishment are remnants of old ideas which are not just outrageously misguided now but were radically stupid in their day. These ideas were nonetheless reified not through a careful process of argumentation but as a result of the consensuses reached by those of a higher social caste whose class-specific experiences and interests, forged in material conditions of privilege, aligned so perfectly they simply assumed they must all have been onto something.

In truth, these proposals, theories and beliefs were merely the expressions of the class interests of those who posited them. The questions they were asking, boiled down, were some variations of 'why can't these poor, sick, vulgar, unhealthy, uneducated, uncivilised people just be great like us?' Implicit in the solutions they proposed were wild overestimations of their own capacities. These false beliefs were then diffused through national institutions of politics, education and media and subsequently adopted by an emergent middle class which came to regard the falsehoods as common sense.

Note how few of the proposals or theories outlined above challenged the existing power structures of their day, nor the class interests which underpinned them. It is then that we begin to understand what 'merit' really means and what 'meritorious' behaviour truly entails: merit is defined to the advantage of the dominant social classes.

The ideas we internalise around demonstrating our merit are not always universal matters of common sense, they are often very specific and are, to varying degrees, about exhibiting a willingness to adopt and adhere to the values, beliefs, attitudes and behaviours of those nearest the top. Do we feel more confident when we dress up in suits because suits are smart or because that's how the rich used to dress? A butler once told me that many moons ago, a king loosened the bottom button of his waistcoat due to a bloated belly and within days his courtiers were doing the same. Why? What was the merit in that? Do we dial in our accents to be understood or does our experience tell us that concealing our regional dialects may produce more favourable outcomes for our careers and social standing? Why should those of us not privileged enough to enjoy an Oxbridge education be expected to modify our speech so as to accommodate the underdeveloped, inferior communication skills of the apparently sophisticated?

And why do you stand a greater chance of going further professionally by saying nothing of these absurdities, perhaps to spare the blushes of your betters whose delicate feathers must not be ruffled, when there is dignity and, indeed, power in recognising and pointing them out?

To be regarded as a person of merit, it is not enough to be decent, honest, hardworking and talented – you must also be willing to integrate those attributes into a social performance. You must be smart enough to sense the absurdity but wise enough to say little of it. If you are unwilling to do so or are not even aware of this caveat, you risk being deemed unworthy of a great many opportunities. When people speak of meritocracy, what they think they are referring to is a country where hard work, talent and integrity are rewarded, when what they are really talking about is a society where opportunity is often distributed on the basis of a person's suitability in preserving a status quo which no longer serves a majority of people.

Meritocracy, as it is referred to and discussed today, is merely the latest in a long production line of half-baked, self-serving theories to do the rounds in Britain. It emerges from a well-meaning desire on the part of elites to create a fairer and more prosperous society – who are constitutionally incapable of factoring their own hubris, ineptitude and self-interest into the equation. The belief that we live in a meritocracy is symptomatic of a collective psychosis. One which causes the sufferer to internalise powerful, intoxicating myths about the extent of their own capabilities while remaining blind to the advantages that permit them to fall effortlessly upward, despite their flaws, failures and mediocrity. Meritocracy is the biggest lie ever told by those who would rule over the rest – not to us, but to themselves.

Not every person from an upper-class background is dangerously naive about what life entails in the lower orders. Indeed, many good ideas, rooted in noble and thoughtful questions, have emerged from the minds of men and women born into privilege. The problem with Britain presently is that we have too many of the other sort. People who evidently possess little in the way of talent, values or integrity but who are nonetheless regarded as worthy as a result of how merit is currently defined, in an era when acting ruthlessly in your own short-term interest is celebrated and seen as a legitimate and even noble way to behave.

Boris Johnson evidently has his charms but surely the moderate ability he so clearly possesses should have seen him rise no higher than the student debating society. He is, by every measurement, an average man, notable only because of the many stupid things he has said and done and somehow gotten away with. Yet he is now one of the most powerful leaders in the Western world. Yes, like all prime ministers, Boris moved into Downing Street as a result of a democratic process. But would he have pulled off the impressive feat had his starting point been Pilton and not Eton? And are we doing ourselves any favours as a society by turning the other cheek, numbed to the evident truth that Boris Johnson's ascent is perhaps an indication that something may be wrong? Is he not a symptom of a deeper malaise? That in Britain we are nowhere near the meritocratic nation he and many of his predecessors have insisted we live in.

Meritocracy is, of course, the highest possible social ideal. But in a society so ravaged by class inequality, it is so rare that people from different social classes even encounter one another, let alone compete in education, employment or business. Therefore, the idea that we all vie for a place in the hierarchy by demonstrating competence is only partly true – we compete in parallel hierarchies based on our social class.

Like many observing the Conservative Party leadership contest in 2019, perhaps worryingly adjusted to the permanent dysfunction on display at Westminster, I found myself strangely detached, comfortable even, as a new and most devious bloom of political invertebrates moved on Number 10 like a swarm of killer jellyfish. Aside from the obvious economic, political and constitutional crises that continue to shape events in the UK, it seems clear that there is a link of sorts between the sharp decline in competence at the top of British politics and the dire situation the country finds itself in. At least when Thatcher set about the unions in the name of free-market fundamentalism based on the now discredited monetarist theories of economist Milton Friedman, she did so with a wild-eyed, orgasmic conviction. Unlike many of the pretenders she undoubtedly inspired to enter politics,

Thatcher was a towering figure, driven by the core belief that cometh the hour, cometh the woman. She truly believed that nobody else at the time possessed either the gravitas, skill or courage to drag the UK economy out of the swamp of high inflation and economic stagnation which had slowly eroded the electorate's faith in the left. Her legacy continues to cast a long shadow over the country she led for 11 years, as well as her many successors, who have either attempted to define themselves – or been defined by others – in relation to her.

Thatcherism, as a political philosophy, was rooted in something. Thatcher had principles and operated from a system of belief. Her legacy will always be hotly contested but no serious person could claim her unlikely rise to the highest office in the land, and the subsequent decade at the top which she enjoyed, came as a result of privilege. Thatcher was an outlier both in terms of gender and social class when she entered politics. She was held to a higher standard than many of her male predecessors because of that. Though it pains me greatly to say it, Margaret Thatcher earned her seat at the top.

Of course, the fact the daughter of a grocer managed to ascend to the very top demonstrates, for some, that anything is possible. On those rarest occasions when unlikely candidates rise, the event is integrated into the national mythos. The national story is revised and the system, dominated by alumni from a handful of elite schools, is re-authenticated as fair. If Thatcher can do it, anyone can. But Thatcher's worthiness lay not simply in her famously intense work ethic and steely determination. She exhibited her merit in her zealous commitment to the emerging monetarist consensus in the United States. And while deindustrialisation was already well under way before she entered Downing Street – Labour had already closed more coal-pits than she would go on to shut, though did so as part of a broader industrial strategy – Thatcher moved the needle where privatisation, deregulation and the corporatisation of public life were concerned. This was how she demonstrated her worthiness. Britain – and her public realm – was put on sale. Thatcher certainly possessed a formidable mix of resilience, talent and competence, but one person's success or failure when discussing the broader issue of inequality of opportunity is irrelevant. In any event, there's a strong argument that Thatcher's talents would still only have gotten her so far – it was her complete fealty to free-market ideology that secured her position. She is the exception that proves the rule.

For those who will never stand a chance of ascending to such lofty heights, the myth of meritocracy operates much like a religious belief. Like class, meritocracy is just a concept. But unlike class, there is little evidence meritocracy exists, despite the fact so many believe it's real. The great irony here also speaks to the confusion at the core of British culture; those who believe we live in a meritocracy tend also to dismiss the existence or relevance of social class. This widely held misconception has led to mass confusion about the true nature of British society, how it is structured and organised.

I recall one gentleman in the *Question Time* audience attempting to argue with a straight face that his £80,000-a-year salary did not place him in the top 5 per cent of earners. Indeed, millions of middle-class people in Britain wrongly identify as working class. They are unsettled by the suggestion that their success and prosperity might be as much to do with the favourable circumstances into which they were born as they are about merit. They reach back into their family histories, desperately trying to locate a distant relative who had type-2 diabetes or cleft lip in order to authenticate themselves in the event the issue of their private education, the bank of mum and dad or the plush postcode they grew up in ever comes up.

This is just one consequence of the promulgation of the myth of meritocracy – people who believe they are working class because their grandad spoke to a miner once. The clue that something is amiss lies in the performative pride they exhibit when talking about their working-class heritage – any genuinely lower-class person knows that ascending the ladder's rungs is in part contingent on concealing traces of your origin. Dialling in your accent. Wearing a blazer instead of a hoody. Claiming to have really enjoyed *Fleabag*. You would only profess to be working class if you were confident that it could have no detrimental effect on your social or professional standing – a confidence which comes with a more privileged upbringing. The other explanation is that class, like everything else today, is construed through the lens of individual identity. Something in the eye of the beholder. Class becomes a subjective interpretation and experience – not a measurable, objective fact.

Oddly, what many middle- and upper-class people who genuinely believe that we live in a meritocracy usually can't tell you is where the term 'meritocracy' originated. In Michael Young's 1958 dystopian tale *The Rise of the Meritocracy*, the narrator describes a society in the year 2033 where social divides are no longer rooted in wealth or poverty but, instead, citizens rise and fall according to their ability. Spoiler alert: the utopia descended into disorder when those regarded as less intelligent eventually revolted. This is because the attitudes, behaviours and values which were deemed meritorious were always defined to the advantage of the winners. The term 'meritocracy' was created for the purposes of social satire but has since been appropriated by politicians, celebrities and business leaders who often benefit from a level of privilege which eludes most people. Meritocracy, as a concept, has been fast-tracked, despite the lack of evidence for its existence, because this idea acts as a worthy substitute for the traditional notions of class which are far more threatening to those who benefit from the status quo. In an ailing social democracy like Britain, the concept of class is as dangerous to the power structure as the telescope was to the idea that the Earth was situated at the centre of the universe.

Meritocracy cuts to the heart of the political crisis in the UK. The ravine that separates the winners and losers. The more privileged your point of origin, the more remote from the reality everyone else experiences you may become. With increasing degrees of affluence, you may lose touch with the truth of your success, because as well as enjoying the great many perks which exist in a society tipped in your favour across health, housing, criminal justice, education and the labour market, you are also furnished with a narrative which conveniently accounts for them. This lack of proximity increases by orders of magnitude the wealthier you happen to be, as does the hubris and the divergence from social reality as it is experienced by those who enjoy no such perks.

The vast disparities on display in society are not accounted for solely by merit. The reason people in poorer communities are likelier to fair less well than those from more affluent ones is because both are, in large part, products of their respective environments. The only way you could test whether meritocracy is real would be to pluck Boris Johnson out of Downing Street, remove his social, economic and cultural advantages and air-drop him into Pilton on a Friday night. From there, he would have to climb every rung of the ladder, competing in every weight-class, drawing only from his innate ability. If he could ascend back to the greatest office of state, without the benefit of wealth, powerful social connections and his dangerous abundance of overconfidence, then I would be first in line to declare him truly worthy. If he couldn't, well, tough luck – that's life.

The myth of meritocracy persists because we all like to feel like we have a degree of autonomy over our circumstances. That we are the authors of our own stories. And like those who rise to positions of great consequence, we are all guilty of embellishing our experiences to some degree, revising the narrative and failing to understand the social forces which determine our trajectories in life. We are all capable of self-delusion. But some delusions are more consequential than others. When meritocracy's most zealous advocates have no frame of reference for genuine struggle and are rarely forced to compete with those outside their own social class, who are they to define what is meritorious? Who are they to dispense advice about matters of personal responsibility? Surely you should have had to overcome great odds yourself before you get to define what or who is and isn't 'worthy'? Indeed, it is the endless falling upward of profoundly unserious and mediocre people in British public life in recent years which exposes meritocracy as bare-faced lie.

Much like a drug addict reaches out for an illicit substance to quiet feelings of discomfort, thus turning down the unbearable volume of reality, popular notions of meritocracy act as an ideological sedative, stilling that nagging sense of imposter syndrome many rightly feel. A quick fix that numbs the user from the fleeting but horrifying thought that their success, prosperity and sense of security may hinge in some way on the struggle of people further down the food chain, and that were the system truly meritorious, they might not find themselves in such favourable circumstances. When your kid's university place is guaranteed because working-class kids' grades are suppressed, your home is worth more because so many people can't own one, and your company can only turn a profit by paying poverty wages, perhaps you should, occasionally, feel like a bit of a chancer. What if imposter syndrome is not always the unfounded suspicion one is undeserving of their success, but sometimes, just their conscience gently reminding them that they really aren't worthy?

If it were only elites suffering from this cultural delusion, perhaps the situation in Britain would not be so grave. Put simply, if all the best people are in all the top jobs, why is Britain such a fucking bin fire?

Interlude
When you don't know what you don't know

Bad science begins with a bad hypothesis – any scientist will tell you that. A question dictates the direction of travel. It sets parameters. What are we trying to find out? Why do we want to know the answer to one question and not another? And who decides why some questions are prioritised over others? In the scientific community, there exist robust methods of establishing these things. Painstaking processes of observation, investigation, testing assumptions and collecting data which is then reviewed by peers for errors.

A particular question can betray our unconscious biases or ignorance. Sometimes that ignorance is innocent. Sometimes it is wilful. Think of the questions a child might ask and what those questions reveal about that child's understanding of the world and reality. Bad questions are not bad in and of themselves. Everyone begins from a place of ignorance. The problems arise in a democracy when bad questions gain a level of traction to the extent that they are taken seriously, despite their ludicrous nature. They become legitimised when people widely regarded as worthy of respect, or as educated, or as informed merely because of how they dress or speak, labour them, repeatedly. I recall one *Question Time* segment in November 2021, which asked: 'Can racism in cricket ever be defined as banter?' I haven't experienced racism so am not at liberty to say, though I suspect the answer is no, but I'm willing to speculate that such frivolous framing of the serious issue of racism is not only irritating and even painful for many who have, but also fundamentally misses the opportunity to dive deeper – workplace racism is widespread in the UK because racial superiority is woven into our national DNA.

Where class is concerned, there is similar ignorance. Much of the discussion around poverty today revolves around similarly terrible questions like, 'Why are the poor so feckless?' Such a question exposes a lack of understanding and could be modified, making it more useful. For example, we could ask, 'Does poverty-related stress impact decision making?' Or, better yet, perhaps we might ponder something like, 'Are those people judging the poor as feckless projecting their own sense of agency and choice onto those who have little agency and whose choices are limited?'

Think of all the wasted opportunities to understand poverty more deeply that are routinely squandered. History is littered with evidence that the assumptions of those of a higher social caste are not only quite often wrong but wrong in ways that are extremely consequential for the rest of us.

This issue of proximity is dangerously underplayed in society and looms over everything. Think of the local authorities, the country over, that profess to be doing all they can to combat stress and destitution, while simultaneously sending out hundreds of threatening letters every day to the poorest citizens, warning of court proceedings for non-payment of council tax. Consider the UK government's commitment to invest more in mental health services, despite rolling out welfare reforms and immigration policies which are increasingly linked to homelessness, food poverty and suicide.

When powerful people misunderstand the true nature of society's problems, yet act confidently based on their assumptions, the damage is felt by the poorest and most vulnerable for generations. How do we close these proximity gaps? How do we bring those with power and influence a little closer to the action? Could we begin devising solutions to social problems based not always on the political requirements of leaders or parties, nor the media-generated bias and ignorance of sections of the public, but in closer collaboration with those who confront these problems every day? I believe we could and, more to the point, should strongly consider bringing so-called 'lived experience' into the social-policy formation process.

*

Now a quick demonstration of what I mean by 'proximity gaps' and how they hamper social policy. Let's imagine I'm pitching you a plotline for a short film about the subtler impacts of low proximity. A fire has broken out in a household in an imaginary working-class community. The fire is being viewed from three perspectives simultaneously. The first, from the perspective of the single mother and three young children who live in the house. The second, from the perspective of a GP who happens to be passing on her way to a house call. The third, from a hillside over a mile away, where a local politician walking his dog notices a plume of smoke rising into an otherwise cloudless sky.

From the vantage point of the mother and children, the fire represents something urgent. Something dangerous and life-threatening that must be attended to immediately. Meanwhile, the GP outside cannot see the flames, only smoke, and can hear the sound of children screaming. She assumes there is a fire and that there are people at home but other than that, she knows very little about either the intensity of the fire or the circumstances that led to it. On the hillside, the local politician takes in the arresting vista of the constituency he has represented for many years, admiring from a distance how pleasant and orderly it appears. He assumes that whatever is happening to cause the smoke on the horizon, someone else will deal with it. While his concern is sincere, it dissipates quickly as he returns to his own thoughts.

A fire engine arrives at the household and the flames are extinguished. The cause of the blaze was a bonfire the mother and kids were having in their back garden. News spreads around the community and the mother and kids are judged harshly by their neighbours. The fire service begins investigating the cause of the blaze. They find remains of metal, plastic and rubbish in the burnt-out yard. They conclude the mother was burning inappropriate materials in her garden and that this was the primary cause of the fire's rapid spread from the garden and into the house.

A few weeks later, the mother attends a GP appointment with her eldest child who is still suffering from the effects of smoke inhalation. Throughout the course of the ten-minute appointment, the mother also discloses that she is suffering from mental health problems brought about by lockdown. That she has a lot of financial anxiety and that her partner is drinking a lot. She is unaware that her GP already knows about the fire and does not mention it out of embarrassment. The GP informs her that she was outside her home during the blaze. That she contacted the fire brigade. The mother is mortified – for her, the incident is now defined not by a near-death experience but by the shame she was made to feel by many in her community for allowing the fire to occur in the first place. The GP tells her it was not her fault. That these things happen. That lockdown has been a tremendous challenge for everyone.

It is only then, when she feels comfortable enough, that the mother begins to open up about the situation which preceded the fire. It emerges that she was burning rubbish because the bin collections had stopped, the dumps were closed down and none of her neighbours would co-hire a skip to get rid of the build-up of excess waste. She was burning rubbish because she had a serous sanitation problem due to a lack of information and coordination at local level with regards to refuse collection.

The GP shows her a public notice with information and soon learns the mother can't read. Upon hearing this, the GP contacts the local politician. It takes weeks to get a response. After exchanges by email and letter, in which the GP outlined that the house fire was partly a result of bins not being collected, dumps being closed and public relations being non-inclusive, the MP promises to do something about it. Two years later, a local law is created which guarantees that in the event of lockdown in future, low-income households with sanitation issues will be guaranteed a pick-up from the local council.

By the time a problem is identified and communicated up the democratic structure, it's often too late. There are so many obstacles, filters and revisions a basic social concern or complaint must go through before it arrives on the desk of someone who can do something about it. Even then, if it does, there are also the varying degrees of urgency, accuracy and fluency with which the problem is regarded, interpreted and understood as it is passed up the chain of command. In some ways, the democratic process renders the simple exchange of actionable information as fraught with the same potential for miscommunication as it would have been had the mother, the GP and the politician all been trying to shout loudly to each other across their relative geographical, social and cultural distances at the time of the fire.

By the time a solution is devised and sent back down in the form of a policy, the conditions on the ground that created the problem may by then have evolved into something else. This is a central dilemma of representative democracy and why so many people feel politicians do not understand or care about the daily lives of their constituents. Politics is often so out-of-sync with what's happening in the poorer pockets of society that even where leaders do actively wish to help, they often hinder instead.

What someone with lived experience possesses that many professionals and institutions do not is arguably the most important component of understanding a social problem and also key to closing that gap – proximity. People with lived experience bring us closer to the action. Why does a tradesman have to look at a blocked toilet or a burst pipe before he lets loose with the tools? Why does a GP need to see you in person and not simply diagnose your symptoms over the phone? Why do news broadcasters send reporters to the scene of the event and not just write up the story based on what they've heard through the grapevine? These conventions exist because it has been established that it would be absurd and even dangerous to do otherwise. Yet decisions about Britain's most pressing social problems in the twenty-first century are often devised remotely, behind closed doors, with no meaningful input from those the decisions affect.

Problems like poverty, while often discussed in broad systemic terms, involve human experiences which must be understood before the issue can be adequately addressed. This is not about getting a handful of people in to 'share their story' – it's about applying the same principle to confronting Britain's social ills as we do to political polling or basic market research. Individual experiences must be surveyed at scale. While all experiences are to some extent unique, and people are famously unreliable narrators, if you ask enough people the right question, common themes will begin to emerge, irrespective of what they say or don't say, and those themes will inform future action.

*

In 1936, three years before the outbreak of the Second World War, the Unemployment Assistance Board commissioned the Pilgrim Trust to make an enquiry into the effects of long-term unemployment on people. The descriptions drawn in part from interviews with the unemployed, found in the National Archive record, are extensive and reveal not simply the lived experiences of the poor and downtrodden but also the learning journey of the middle- and upper-class professionals surveying them. Professionals not only struck by the abysmal conditions of their lives but by specific details of their experiences with respect to a then threadbare and inadequate system of social protections:

'We have spoken above about the economic cycle in working class life. There is a much shorter economic cycle, that of the week. The man on unemployment assistance ordinarily draws his money on Thursday. "The day you get your money and the day after are the only days you can get any relish," said an old man living a Liverpool cellar, "otherwise it is just bread and butter." A Liverpool housewife said, "We have a good blow out on Thursday and Friday and Saturday, then go short for the rest of the week"; and a woman in Leicester, "After Sunday I often don't know what I am going to do for the rest of the week." What happens on Monday in pawnshop ridden neighbourhoods is that Sunday suits and any other likely articles go into the pawn shop, to be redeemed again, if they are lucky, on Saturday. This shows the narrowness of the margin on which such families are living. There are not a few pennies over at the end of the week, but a few pennies short. It is a commonplace that there is all the difference between a household managed well and a household managed badly, and that in some homes the allowance goes further than in others. But even in the homes that are best managed, indeed particularly in such homes, we cannot but be aware how delicate is the balance. One of the investigators visited a young couple in Liverpool, aged 26 and 23, on a bitter February afternoon. It was snowing outside. The house could hardly have been better kept and both of them were neatly dressed. Yet there was no fire, and so far that day – it was three o'clock – they had had nothing to eat. They lit the fire when he came in, for the man said "his mother had just helped them out with a bit of coal", so they could manage it. He said his wife "had something for this evening, and that they weren't starved, though sometimes they do go pretty short". It is a household like that which shows how difficult is life on the dole, however careful the housewife may be.'

Contained in this passage are three key insights that any unemployed person would have known intuitively, but for the professionals conducting the investigation, these experiences were so striking as to be worthy of note. The first, an insight that poor people lived week-to-week. The second, that households often resorted to pawning their valuables for money to obtain essential items like food. And the third, that this weekly cycle of hardship was invariably the case for most households, irrespective of whether they were run responsibly or not.

Over 500 such interviews were carried out across the UK, each illuminating specific aspects of the lower-class experience. These would later inform the Beveridge Report which, after the war, led to the creation of the Welfare State – the most radical social reform in British history and one which transformed society. Within 20 years, poverty fell as the UK reached near full employment. Slums were cleared and housing estates and schemes were built, complete with baths and toilets. The creation of a free education system and health service contributed to a reduction in infant mortality, an increase in average life expectancy, wider social mobility, rising household incomes and a sharp decline in the inequalities across health, education and employment which had, until then, typified the British class system.

Despite the very obvious logistical complications of opening social policy formation and analysis up to scrutiny in this way, when surveyed and understood correctly, lived experiences provide valuable insight. We begin to comprehend what we are dealing with at a level that was previously unattainable. There is a great deal of difference between drawing on a handful of carefully selected lived experiences (for political purposes) and drawing on hundreds or even thousands in good faith. Ultimately, it's about the quality of the question. If the motive is to validate prior assumptions or political expediency, lived experience, however gratifying to hear, is not terribly useful. It becomes simply another means by which value is extracted from the poor and transferred to the affluent. But if the intention is to view, in higher resolution, a social trend such as food poverty, or rough sleeping, or drug addiction, then lived experiences may provide insight, supplementing existing data thus creating new possibilities for meaningful action and change.

There are few substitutes for it. And while the debate around the utility of lived experience rages, suspiciously little work has been done to ascertain the risks posed by its conspicuous absence. The risk when individuals and institutions, presiding over social problems, possess little or no experience of what it's like to live with those problems. Then again, the evidence that a lack of lived experience on the part of those in positions of power and authority may create as many problems as it solves is all around us.

ACT TWO

Fucked Left,
Right and Centre

15

Putting the House in Order
Poverty, political dysfunction and parliamentary privilege

'I'm not asking the public to trust me. I'm asking the public to trust themselves. I'm asking the British public to take back control of our destiny from those organisations which are distant, unaccountable and elitist and don't have their own interests at heart.'

Michael Gove (Secretary of State for Levelling Up, Housing and Communities), Sky News, 2016

In 2021, Conservative MP Owen Paterson was found to have made several approaches to government departments on behalf of two private companies that were paying him more than three times his £82,000-a-year parliamentary salary. Facing a 30-day suspension for his egregious breach of lobbying rules, the UK government circled the wagons, not to hold one of their own to the same standards they had been holding people on benefits, but in an attempt to rip up the rule-book entirely and spare their man the consequences of his actions.

In an exhibition of pure, undiluted privilege and entitlement, they sought to simply change the rules because the rules didn't suit them and, in the process, give a firm twos-up to whatever expectation remained of basic ethical norms in Westminster. By attempting to place Paterson above the rules, the Conservative Party was prepared to set merely the latest in a long line of worrying precedents which point to a vertical freefall in parliamentary standards and a dangerous accountability gap at the heart of British democracy.

Within hours, a U-turn followed in the face of a justified public backlash, aided in part by a reluctant right-wing press. Stories of similar breaches hit the news, as well as examples of MPs drawing rather generously from the public trough. MPs, who rage at the thought of single mothers, unemployed people or migrants 'wasting' public money or 'abusing' the benefits system, had been taking generous government subsidies for homes and then renting them out. MPs had forgotten to declare earnings from their second or sometimes third jobs to Parliament. Suddenly, as talk of banning second jobs swept the country, many of the same MPs who had only weeks before voted to reduce Universal Credit payments by £80 per month for the poorest families were lining up to go on television and proclaim that banning second jobs was tantamount to cruelty and would hit MPs lifestyles hard. MPs are human, too, you see. They have families to look after and that we should consider carefully the prospect of compelling them to live on a measly £82,000 salary, as such low pay might lead to a decline in quality in politics.

That tough talk about following the rules or paying a price which had been so punitively applied to people on benefits the previous decade was suddenly, when potentially applied to them, a really scary prospect. How must they be expected to live on three times the average salary, getting only their shopping, gas, electricity, taxis, trains, flights, Pimm's, paperclips and petrol covered by taxpayers? Was it any wonder so many of them were forced into the humiliating practice of renting out one of their many dwellings, thus drawing an essential, life-saving additional income, while the state subsidised their accommodation costs?

I know some of you reading this have repeatedly found yourselves on the receiving end of the harshest reforms the Welfare State has ever seen but spare a thought for the politicians who voted them through. Try to see it from their point of view, won't you? Imagine how it would feel if you only earned enough money to have three cars. If you only had two homes. If you couldn't claim the 78p fuel cost of a two-minute drive to the post office back as an expense. Imagine having to occasionally stay in a downmarket £200-a-night hotel with no mini-bar, complimentary umbrellas or sultry mood-lighting configurations. How would *you* feel if you could only afford to send one academically mediocre child to a fee-paying school, rather than all six? That was the bleak abyss into which some of these poor, unfortunate, multi-millionaire public servants were forced to stare. And to think they were plunged into such a nightmare because Boris Johnson sought to leverage the power of government not to help the poorest but to protect one man from a 30-day suspension – a pitiful punishment for lobbying in the first place.

Upon his resignation, Owen Paterson spoke of the 'cruel world of politics'. Cruel indeed. Though perhaps a night at one of Britain's immigration detention centres may have caused him to choose his words more carefully. This from a man who, according to They Work For You, a site which tracks the voting records of MPs, voted to reduce benefits to the poorest, including to disabled people, and who endorsed repeatedly the harsh system of rules and penalties which have immiserated some of Britain's most vulnerable citizens – while using his privileged position in public office to enrich himself.

If reliving this hypocrisy makes you angry, however, please be sure to mind your tone. No mention of hurtful words like 'corruption' or 'scum', please. Many in the party of law and order, who believe deeply in the notion of meritocracy and personal responsibility, not only expect to play by a different set of rules than the rest of us but are also, apparently, of such brittle spirit that as well as struggling to live by the standards they demand of those who have least, they also find the mere thought of offensive language deeply traumatising. That's why you can't call them liars in Parliament – even when they are being dishonest. And that's why they must be referred to as 'honourable' – even if they are nothing of the sort. Sadly, however, the very clear lines of political attack which I outline here, which could have been pursued by other parties with respect to the fundamental unfairness at the very core of our political system, were not open to opposition leaders – the culture of entitlement is so endemic that the contagion from the inevitable scandal, across all parties, had a Starmer or a Sturgeon truly gone for the jugular, would have made the expenses scandal of 2009 look like a socially distanced walk in the park.

A sceptic might argue there exists one set of soft-touch, opt-in rules for some and another set of absolute, punitive and potentially deadly ones for others. An ethicist might argue that there is no moral difference between making a fraudulent benefit claim – which can see you left to starve – and forgetting to declare £30,000 in earnings or £6 million in loans – which will get you a playful slap on the wrist. An optimist might claim that politicians suddenly remembering to declare income from second and third jobs before kindly referring themselves to the relevant parliamentary standards committees was proof positive their oversights were sincere. My point is simple: who cares if they were sincere? Nobody at the Home Office wants to know if the failure to produce four pieces of documentation is sincere – you get deported back to a warzone. Nobody at the DWP wants to know if your being ten minutes late for a compliance meeting was sincere – you are sanctioned and forced to use a food bank for six weeks. Why do so few people who regard themselves as intelligent fail to grasp how grotesquely unfair this is? And for anyone who feels I'm going a little harder on the Tories right now, that's because they've gone so hard on the voiceless, the weak and the infirm. Indeed, I would go even further, and so callously contend that the selective application of rules and penalties as they pertain to politicians as opposed to their application to the ordinary people upon which many of them have waged an open war of institutional hostility is no better than a form of organised crime – so imperceptible to the both the perpetrators and the victims that it feels strangely normal.

In Parliament, appearances are everything. There's the building itself, which is quite simply spectacular. Looking on at the grand, gothic 300-foot riverfront façade from a boat on the Thames induces a dreamy reverence. The Palace of Westminster, which began construction in 1840 and took 30 years to complete, was painstakingly designed by the finest artists money could buy to project an unassailable image of God-given authority. Once you're in, however, it looks and feels more like a dusty old museum than the cockpit of an advanced Western democracy.

So much of the pomp and ceremony associated with Westminster is about projecting honourability. This extends to how parliamentarians are permitted to address one another in the chamber. MPs can be expelled for rudeness. This strict etiquette is designed to deter undignified and heated exchanges.

Labour veteran Dennis Skinner was famously invited to leave the chamber for referring to David Cameron as 'dodgy Dave', a prescient reference to the then-prime minister's generous usage of his parliamentary housing allowance. Skinner wasn't wrong but his behaviour was deemed so dishonourable that speaker John Bercow had no alternative but to boot him out for the day.

I attended PMQs in 2019, keen to witness the spectacle up close, after years of shouting at my television 800 miles away. Despite great efforts to make Parliament more accessible to the public, being sat there felt rather awkward. I was pulled up for impropriety on two occasions before PMQs even started – once for being dressed too casually and once for trying to take a picture from the press gallery. Despite every MP in attendance being sent there to represent people like me, I experienced the faintest sense that people like me weren't supposed to physically turn up. That most citizens do their bit for democracy by keeping their distance. The business of running the country, it would seem, must always fall to the same kinds of people. The right people.

That so much of the responsibility for the UK's democratic decline lies at the door of politicians who cannot detect just how remote they are from reality renders antiquated parliamentary procedures, designed to project honourability and prestige, all the more pitiful. Indeed, that this institution, so fortified from the desperate socioeconomic conditions it has created in many parts of the country, is so evidently failing, is about the only thing most people in Britain can now agree on.

In Act Two, having established that class inequality in Britain is both structural and cultural, and that the central problem of our time lies in the dangerous distance between government and the governed, we must now consider the role political ideologies found across the spectrum have played. Why have successive Conservative governments returned time and time again to the politics of hostility with respect to the poor and vulnerable? Are they evil? Are they stupid? Or is some aspect of social reality as it is experienced by the less fortunate, often on the receiving end of these relentless waves of institutional and cultural malevolence, getting lost in translation? Why, in an age of such undeniable inequality, where the average worker would have to labour for millions of years to amass anything close to the wealth of some individuals, has the Left (the actual Left, not liberals with a guilty conscience) seen such little cut-through with working-class people? Is a hostile right-wing press the cause? Are working people simply too brainwashed by capitalism? Or has the radical Left simply been on the political margins for so long that it's lost contact with the notion that winning power requires more than moral certitude and grassroots organising? Why did the populism of Nigel Farage, which played arguably the most decisive role in the UK's tumultuous exit from the European Union, gain so much traction in the UK when it was so evidently riddled with falsehoods and hyperbole? Was it his wealthy backers that gave him the edge? Was he just a bit-player in a vast global conspiracy to upend social democracy? Or did he speak, however vaguely and dishonestly, to a justified sense grievance felt by millions who haven't been cut into the action of globalisation? And what role do the self-styled fair-minded moderates in the hallowed centre ground hope to play? Is Britain's predicament, as so many of them seem to think, an unfortunate result of the vacuum left by their political absence or can its roots be traced directly back to the golden age of centrism, Tony Blair and the New Labour delusion that the class war was over?

While recent years have seen the rise of a narrative which attempts to partition the Brexit referendum and everything that followed, including the ascent of Boris Johnson and the UK's terrible handling of the pandemic, from a simpler time in the not-too-distant past, when politics was serious and politicians were nobler, I'm afraid that comforting lullaby does not stand up to scrutiny. Parliament's inability to truly concede the role it has played in Britain's dysfunction is nothing new. This disconnect between many in public office and the ordinary people on the receiving end of their collective efforts is historic. For as long as there has been a democracy to participate in, working-class people and the poor have had to organise and fight for every right, every liberty and every other socioeconomic advance they got – never forget that.

While social progress in areas like worker's rights, children's rights, racial and gender equality and sexual expression are so often cited as central achievements of a nation's parliament, or even of individual politicians, before being integrated into the British national mythos, in truth, government has all too often been the last remaining impediment to social and economic justice. Working-class and women's rights, the abolition of slavery and child protection legislation are often held up as the achievements of politicians or institutions but the British establishment resisted all of them, often violently, either to guard the interests of those who benefited from the injustice or because they thought they knew better. Radicals were hunted down and hanged as traitors in the street. Slave owners – coincidentally, members of the same social caste as many of the politicians – were compensated handsomely for the inconvenience caused by abolition. Meanwhile, women were last to be granted their rights and liberties; vilified, ridiculed, beaten down by police and even killed, women were met with the full brunt of state coercion and violence as they rose up to wrest basic democratic dignities from a patriarchal political class which could not see that a democracy where a majority of the population was female, but where women could not vote nor stand for election, was no democracy at all.

Now, well into a twenty-first century which promised so much yet has delivered so little, we see a backslide into the same problems as before, only this time Parliament has a new multicultural veneer to conceal the root cause of a slow erosion in many of the advances so often cited as evidence of Britain's social, cultural and moral superiority: the dominance of one social class. Specifically, a decline in working-class MPs and the rise of middle-class career politicians. The Commons has more women, LGBT, and black and ethnic minority MPs than ever before but working-class voices are conspicuously absent.

Research published in the journal *Comparative Political Studies* examined the policy preferences of working-class and career politicians within the Labour Party (regarded by many as the primary vehicle to advance the interests of working-class people) both before and throughout Tony Blair's leadership. The study revealed that MPs from working-class backgrounds tended to favour traditional welfare policies more than their careerist colleagues. During the Blair era (1997 to 2007), there was a substantial drop in working-class politicians and, in the same period, a decisive shift towards more centrist polices around reforming welfare. The report found that career MPs – politicians that come from a background in politics or a closely related profession – were more likely to adopt policies for reasons of political strategy, the ultimate aim being to entice enough swing voters from the right of the political spectrum to win elections. Working-class MPs, however – categorised as politicians with backgrounds in manual and unskilled labour – were likelier to support policies that benefit working-class communities.

Study author Dr Tom O'Grady, lecturer in quantitative political science at University College London, in an interview with the university's magazine *UCL News* in 2018, observed that 'political parties across the developed world, particularly European social democratic parties, once consisted of politicians drawn from a broad range of classes and occupations, including manual trades,' but that today, 'many political parties are dominated by middle-class professional politicians with little experience outside of politics itself . . . Working-class people find it increasingly difficult to enter politics.'

One might naturally assume this trend is a by-product of ten years of Tory government but in truth, it began with a party that exists, apparently, to protect the working class. 'Before Tony Blair came to power there was only a modest difference in working-class and careerists positions on welfare reform,' O'Grady said. 'But our research finds that during his premiership – the influence of working-class MPs dropped while there was a rise in the influence of careerist politicians.' O'Grady found that MPs from working-class backgrounds have 'a stronger ideological attachment to welfare provision because it benefits working-class voters,' whereas their middle-class counterparts' 'greater concerns for electoral success and career advancement meant they were more likely to support welfare reforms.' The findings suggest that the large shift from working-class MPs to career politicians in the British Labour Party led to an erosion of representation of working-class voters' interests. 'Put bluntly,' he told UCL's magazine, 'careerist MPs are much more likely to blow with the political winds.'

In the 1920s, MPs from working-class backgrounds made up 70 per cent of Labour's parliamentary cohort. But since the eighties, this number has declined substantially. Today, despite embattled former leader Jeremy Corbyn's populist pretensions, just 8 per cent of Labour MPs hail from working-class backgrounds while he, himself, was privately educated.

Just 7 per cent of children go to private school in the UK but 48 per cent of Conservative MPs, 17 per cent of Labour MPs and 14 per cent of Lib Dems MPs were privately educated. The average cost for a place at a private school in the UK is £16,119 a year. St Paul's (which George Osborne attended) and Westminster (attended by Nick Clegg) cost £23,481 and £28,200 respectively for day students. Compare this to how much most people earn in the UK and you can see how exclusionary this is. Boris Johnson was the twentieth of 55 prime ministers to have attended Eton. Attendance at Eton currently costs £37,062 a year. Only nine (or 17 per cent of) prime ministers have been educated at non fee-paying schools and many of those were selective grammar schools. Both Jeremy Corbyn and Theresa May attended grammar school. Less than 1 per cent of the population study at Oxford or Cambridge universities, compared to 75 per cent of all the UK's prime ministers. But education is just one of many areas where the political class diverges greatly from the wider population.

MPs earn more than the bottom 90 per cent of the country. Less than 2 per cent of UK adults are millionaires, while at one point two thirds of David Cameron's cabinet were millionaires. Theresa May's cabinet was similarly unrepresentative. Only 2 per cent of the UK population are landlords, while in 2016, 39 per cent of Tory MPs, 26 per cent of Scottish National Party MPs and 22 per cent of Labour MPs were landlords. From 1979 to the present day, not less than 10 per cent of MPs from the three main parties have worked as barristers or solicitors (while 0.22 per cent of the UK population are solicitors), and the percentage of MPs that had been publishers or journalists has never fallen below 6 per cent (less than 1 per cent of the population are journalists). Over the same period, the percentage of MPs who were manual workers – such as miners – has decreased steadily from 15.8 per cent 35 years ago to just 4 per cent today. When polled, a majority of the public said it wanted fewer lawyers and reporters as MPs, and more doctors, scientists, factory workers, economists and teachers.

Why does any of this matter? Why should we be concerned that the individuals who tend to rise to govern our society are so unrepresentative of the rest of the country? Why can't we just accept that some people do better than others? That this is simply the way it is. I often ask myself these questions too. It's far easier to sigh in resignation at the growing gulf separating the winners and losers. The haves and the have nots. The powerful and the powerless. Or better yet, to entertain the popular myth of meritocracy which appears to perfectly explain it. If the situation were not so grave, I could understand why so many people choose to keep their heads buried firmly in the sand. Unfortunately, however, this is not a problem we as a society have the luxury of being able to avoid. Or to postpone confronting. We need only look at the policy which has defined Britain over the last decade or so for clues as to how socially inept Westminster has become under the clumsy stewardship of delusional careerists.

At the height of the 2008 financial crash, the Labour government was forced to recapitalise a collapsing banking system – with a rescue package totalling around £850 billion – to prevent what then-Chancellor Alistair Darling believed was the imminent 'breakdown in law and order'. An economic recession soon followed and to cover its outgoings, the government was forced to borrow a record £154 billion in 2009. Chancellor of the Exchequer George Osborne, only weeks in post after the general election that put the Conservative Party (in coalition with the Liberal Democrats) back in power, announced an austerity package of tax rises and public spending cuts in his June 2010 budget. In his October Spending Review later that year, Osborne declared, 'Today is the day Britain steps back from the brink' – and proceeded to announce unprecedented public spending cuts that the *Financial Times* called 'the most drastic in living memory'.

Cuts in local authority budgets, welfare, housing, prisons and policing, which outstripped measures taken by most other advanced economies following the crisis, were sold to the British public as simply unavoidable. Soon after, the harshest reforms in the history of the Welfare State were passed with relative ease. Initially intended to harmonise a complex system of social entitlements, utilise digital technology to increase accessibility and, crucially, to remove the stigma attached to various benefits, the government's flagship welfare reform of Universal Credit immediately plunged thousands into poverty and crisis. Rather than modernise welfare for the twenty-first century, the reform went on to roll back decades of social progress, driving residential instability, homelessness, food poverty, a drug crisis and an increase in suicides. And then, the government, beginning to feel the social strain of its harsh programme of austerity, made the decision to capitalise politically, placating increasing concerns about immigration by passing successive Bills targeting foreign nationals with the most malign, mean-spirited and racist reforms in living memory.

In May 2019, a respected research institute, following a five-year study, felt compelled to create a new metric to track an emerging and worrying trend. The Institute for Fiscal Studies (IFS) used the ominous term 'deaths of despair' – fatalities caused by suicide, drug-related deaths and alcohol-related liver failure. The IFS Deaton Review warned that a lack of employment opportunities, social isolation and relationship breakdown 'may slowly be taking their toll on people's mental and physical health' and that this trend had 'contributed to a small rise in middle-age mortality overall in the last few years, bringing to an end decades of continual improvement'.

In the same week, the UK government's own Social Mobility Commission (the one that didn't resign in disgust) found that 'being born privileged in Britain means that you are likely to remain privileged' while being born disadvantaged means that 'you will have to overcome a series of barriers to ensure that you and your children are not stuck in the same trap.' The impact of social inequality, not merely on those who experience it but on social cohesion in Britain as a whole, has been dangerously underestimated by our leaders, all of whom are largely insulated from its social effects. They have, however, paid for it politically at different points, whether at the ballot box or in the justifiably high levels of scepticism with which the political class is widely regarded.

The following July, a report titled 'Who Governs Britain?', produced by Matt Warman MP, cast new light on the gaping ravine between the British public and the dysfunctional political system presiding over it. The report found that a maximum of only 9 per cent of

Britons placed 'significant faith in any layer of government to do the right thing by them'.

Each of these studies was presented entirely separate from each other. Each looked at one distinct aspect of the crisis. The IFS report focused on health and mortality; the Social Mobility Commission on class disparity, while 'Who Governs Briton?' looked at trust in the political system. However, only when taken together are the implications of each report, and the scale of the task facing the United Kingdom as a whole, truly revealed.

The evidence is in, folks, and could not be any clearer. Many of the metrics by which we have traditionally gauged the health of British society, thus justifying the political and economic systems that underwrite it, and sustaining the careers of those who manage it, from GDP to life expectancy, child mortality, mental health and wellbeing, upward social mobility, violence, crime and the suicide rate, are flashing red.

Driving this stagnation is income inequality, higher in Britain than any other advanced economy except the United States. The ludicrous notion, peddled for decades, that Britain is a meritocratic society where hard work and playing by the rules will get you somewhere has been revealed as either wilful delusion or bare-faced lie. A claim so demonstrably false, yet stubbornly held to be true by so many, that it meets the necessary criteria as to be classed as a superstition. Many of the pressing issues that face ordinary people in Britain remain worryingly theoretical and dangerously abstract to politicians, locked in their political bubble.

The tidal wave of social problems racing towards all of us because of this unsustainable and toxic inequality has the potential to overwhelm society. The incremental nature of the country's social and economic decline presents the comforting illusion that the turbulence is temporary. That it is manageable and can be averted. This mirage emerges in the reality vacuum that exists between a political establishment incapable of correctly interpreting the true scale of the crisis engulfing it and a mainstream media apparatus that has failed to hold it to account on anything but superficial grounds.

Veteran journalist and Channel 4 news anchor Jon Snow used his

MacTaggart Lecture, in November 2017, to warn his profession of a worrying disconnect between the people living in communities and the people reporting on them. Snow said frankly that media was 'comfortably with the elite, with little awareness, contact or connection with those not of the elite' and that the Grenfell fire proved this lack of connection was 'dangerous'.

Just as local people in west London were frozen out of the decision-making processes that affected their lives, millions of Britons the country over have only a tokenistic say over the society in which they live. The forces that shape our national crisis move in from the margins of society, gaining momentum, while many in the largely fortified centre-ground struggle to correctly interpret the nature of the tidal wave racing towards them. Away from the glaring mediocrity of Westminster, the London media bubble that largely functions as its PR company and the middle-class enclaves dotted around the UK which act as productions lines, assembling the gatekeepers of tomorrow, most people in Britain are livid, living in post-industrial landfills that have been regenerated more times than Dr Who.

While society remains distracted by a culture war – the now go-to folk devil for the right-of-centre ruling class, often stage-managed by them and their supporters for maximum distraction – we may have missed a trick by not discussing the risks posed to democracy by a managerial class that possesses little to no experience of what living on the breadline in Britain is actually like. The results of this dangerous lop-sidedness, where the interests of the dominant social classes have supplanted those near the bottom, are not only clear but terrifying.

If Darling bailed out the banks because he feared social breakdown in 2008, after a decade of economic growth and rising prosperity, how will Britain fare in the aftermath of a pandemic which followed ten years of brutal austerity, escalating cultural tensions and political dysfunction? Sadly, too many who undoubtedly entered politics to do good have been subsumed by institutions which are designed to preserve entrenched inequalities and have grown dangerously detached from social reality. It is deeply ironic that a public institution where honour and good manners are of such tremendous import has become the chief incubator of everything which is truly vulgar about Britain.

16

The Right
Conservative estimates – what Tories get wrong and why

In her memoir, *The Downing Street Years*, Margaret Thatcher wrote briefly of the social discord she witnessed when confronted by a run-down housing estate in Liverpool in 1981. It is a brief insight into her thinking where social deprivation was concerned. Revealing that, for all her undeniable skill as a politician, Thatcher, like all of us, operated from a place of ignorance on many issues.

> 'Driving through Toxteth, the scene of the disturbances, I observed that for all that was said about deprivation, the housing there was by no means the worst in the city. I had been told that some of the young people involved got in trouble through boredom and not having enough to do. But you only had to look at the grounds around these houses, with the grass untended, some of it almost waist high and the litter, to see that this was a false analysis. They had plenty of constructive things to do if they wanted. What was clearly lacking, was a sense of pride and personal responsibility. Something which the state can easily remove but almost never give back.'

Thatcher no doubt regarded her observation as deeply original and incisive. Thankfully, greater minds than Thatcher's were, by then, already grappling with the conundrum of why some communities present as more run-down than others, giving rise to crime and anti-social behaviour. In 1982, an article appeared in the *Atlantic Monthly*, in the United States, floating a far more credible theory to explain the phenomenon than Thatcher's rather pedestrian observation. The authors of the piece, titled 'Broken Windows', criminologists James Wilson and George Kelling said, of their theory:

'Consider a building with a few broken windows. If the windows are not repaired, the tendency is for vandals to break a few more windows. Eventually, they may even break into the building, and if it's unoccupied, perhaps become squatters or light fires inside. Or consider a pavement. Some litter accumulates. Soon, more litter accumulates. Eventually, people even start leaving bags of refuse from take-out restaurants there or even break into cars.'

What Wilson and Kelling were describing were the inevitable results of urban environments in social and economic decline, and how the creeping dereliction and lawlessness that arise often beget themselves. What Thatcher so plainly failed to understand was that the unsightly conditions she witnessed were related directly to the radical, sweeping economic reforms she had implemented. While many did prosper under this new economic settlement, others became dislocated. When these social connections were fractured, so too were the informal social customs that underpinned the sense of order and, as unemployment rose, and the spectre of addiction and ill-health loomed large on the horizon, the aesthetic and social decline of great swathes of Britain became a foregone conclusion. In this context, cutting your grass begins to seem rather pointless.

While the broken-window theory was an attempt to account for the rising crime rates in areas of social and economic deprivation, it nonetheless lent context to the increasingly derelict appearance for which these communities would become known. Thatcher's assumption, born of her own limited social experience, became a linchpin of her general outlook on poverty and social deprivation. It was perhaps ironic the Iron Lady was so struck by the sight of unkempt gardens in Liverpool given that the country she had by then been running for two years was, in 1981, growing wildly out of control.

The so-called 'Summer of Discontent', during which a series of riots swept some of the UK's major cities, came at a time when Thatcher's attitude towards working-class resistance in communities, and the networks and institutions through which working-class interests were represented, such as trade unions, had begun to harden. Her visit to Liverpool came just months after the Brixton riots, an uprising born of the growing discontent brewing in post-industrial Britain which found expression along lines of race as well as social class. Beginning on 10 April that year, when a group of young Black men, antagonised by the 943 stop-and-searches carried out in the area over the previous five days, understandably interpreted two white police officers attempting to help an injured young Black male as harassment and attacked the vehicle. The conflict quickly escalated and led to a swift build-up of police. By Sunday, 12 April, a total of 7,000 police officers had been involved in dealing with the disturbances, arresting a total of 282 people, most of whom were young people of colour.

As so-called 'copy-cat riots' broke out over the subsequent weeks and months, Thatcher received requests from senior police chiefs for enhanced powers of arrest as well as more advanced, military-grade crowd-suppressant equipment, including riot shields, longer truncheons, rubber bullets and water cannons. These, it was believed, would aid law enforcement in restoring order and respect for the rule of law. Thatcher's reverence for police and military was such that any requests made were often granted without question. In July of 1981, riots erupted in Liverpool's Toxteth area, followed in the same week by a wave of disturbances in London, Leeds and other cities. By the time Thatcher arrived in Liverpool, the events in Brixton were fresh in her mind.

The key insight that drove Thatcher's social policy was that aspiration was what communities were being deprived of and that this was something they were doing to themselves. On one level, she was correct. However, what Thatcher failed to observe was the effect of the monetarist polices she had adopted to tackle the UK's pressing economic problems. Her failure to grasp the impact of the economic transition from a command economy to a free market on those who could not insulate themselves nor absorb the fallout was arguably the fundamental miscalculation of her tenure. It was certainly the most consequential – we are still coming to terms with the blowback today.

The depth of her misunderstanding extended to the root cause of the riots themselves. This apparent tendency for social unrest and disorder was, in her view, the result of a combination of factors. A lack of pride and the absence of a sense of personal responsibility, coupled with the decline of the church. Indeed, what shocked Thatcher more than anything was how the police, in her view, had been victimised by 'the mob'. Even though it was later found, in the report by Lord Scarman into the Brixton riots, that the police response, and the force's conduct in such communities generally, was central to the severity of the disturbances.

Thatcher's analysis of both the riots and the social conditions that gave rise to them was mistaken on many levels. Her reverence for police, her limited social experience as the daughter of a middle-class grocer and her position as the most powerful person in the country converged, predisposing her to misinterpret the nature of the problem. Her low proximity to matters of race and class was laid bare. In historical analysis, these finer points of a leader's mistakes are rarely noted; Thatcher's tenure is defined by her economic radicalism. But had she not reaffirmed those traditional Conservative values of personal responsibility, law and order, and patriotism so unequivocally, the economic transformation she spearheaded may never have gotten off the ground. Her socially and culturally conservative British nationalist demeanour was, in many ways, a Trojan horse which created the political wriggle-room necessary to undersell her more radical, economically risky ambitions – a pattern Johnson's Tories would mirror decades later by appealing to working-class cultural anxieties when, in truth, the roots of their problems are economic. Thatcher's Conservative government wasn't evil, as is often claimed, but rather an entitled political coterie who believed, consciously or not, that their limited class perspective constituted a complete and inclusive social vision.

Tories are complicated creatures. To understand the devastating impact of Conservative Party social policies, particularly around welfare, one must first understand something of conservativism – not as a political ideology but as a value system. The greatest misconception about Tories is that their primary interest is themselves. Nothing could be further from the truth. While millions could never countenance such a notion, the point remains that while Tories are regarded widely as uncaring, and often regard themselves as realists, conservativism is deeply and ridiculously idealistic.

At its core, conservativism is about tradition. The belief that while change is inevitable, its pace should be moderate. The status quo has been tested and refined over time and so, while imperfect, it is believed to be worth preserving. It is from this core principle that many other conservative values around law and order, family and personal liberty spring. In recent times, traditional One Nation conservatism, which envisioned the state as playing a more paternal role in people's lives, has since developed into what has been termed the New Right, which eschews the post-war consensus and regards the individual as the bulwark of society and free markets as a liberating force.

The problem in Britain today is not conservatism as such (though even its more benevolent forms can be deeply harmful when applied to the wrong context) but its capture by a succession of figures so incompetent, socially inexperienced yet excessively and unjustifiably confident, that even where they desire to do good, they instead create social chaos and political mayhem. This problem is undoubtedly exacerbated by powerful sections of the press, which we explored previously, who share the New Right's same basic vision of society: a low-wage, low-tax economy where immigrants are regarded with scepticism, welfare is seen as an affront to free markets and hierarchy and hereditary privilege are regarded as natural – wrapped in a lovely big Union Jack.

What successive Conservative administrations have done, where social policy is concerned, is demonstrate consistently and irrefutably that they misunderstand Britain's lower orders so dangerously that where there were countless opportunities to bring hope, they have instead brought wave after wave of despair, dysfunction and death crashing down on our most challenged individuals, families and communities. What is most shocking (and perhaps surprising for some) is that this harm often begins as an intention to do good.

Understanding conservatism must be viewed by those who do not subscribe to it as central to pushing back against its worst excesses; whether you like it or not, there are millions of working-class people in this country who either lend their vote to the Tories or have supported them all their lives. Any meaningful long-term action on the broader issue of social inequality that extends beyond electoral cycles requires a firm grasp of the conservative moral worldview. I'm not sure if you've noticed, but calling people 'scum' on Twitter doesn't seem to be having much of an impact – the Conservative Party is the most successful political party in the world and has been handing lefties and liberals there electoral arses on silver platters for more than two centuries.

To understand conservatism, one must talk and listen to, as well as study, those who hold conservative convictions. In doing so, you learn a lot. I have learned, for example, that 'conservatism' functions more as an umbrella term; conservatism does not automatically equate to the Conservative Party but to various combinations of beliefs, values and political ideas which are held with varying degrees of zeal. Some conservatives are driven by a sense of patriotism. Others by religion. Some are socially conservative where others are liberal. We have old-school Tories and neo-Cons. We have classical liberals and hardcore Christians. Even people who regard themselves as liberal or left-leaning likely hold certain views, or have adopted certain attitudes and behaviours, which are, essentially, conservative. The day I became a dad was the day I understood that my instinct to protect my son would, if push came to shove, overpower any sense of obligation I had to anyone else – or to their children. While I do not pretend that it's easy to sit through someone explaining why they would happily outlaw abortion, on the basis of protecting the sanctity of life, while supporting social policies which will later place that same young life at risk, I do believe anyone interested in social justice must, as a rule, be able to temporarily pace this ideological terrain.

Four central themes are fundamental to the average conservative outlook: family, law and order, personal liberty and self-reliance. Conservatives tend to regard the institution of family as central to the development and sustainability of a prosperous society – therefore, the biggest contribution you can make to your community and wider society is to raise productive and responsible children. Successive Tory governments have acted to incentivise marriage and to disincentivise marital break-up for this reason. In the tax system, and the Welfare State, we see numerous policies which are primarily designed to encourage couples to have children, marry and remain together. As society has moved away from traditional notions of family and embraced less conventional family structures, such as single-parent households, same-sex parents, foster and kinship-care models, many conservatives attribute signs of social decline, such as youth crime and drug deaths, not to economic or social inequalities arising from their own policies, but of the abandonment of family values.

They are in a sense correct – it probably does bode better for children to be raised by two parents or caregivers – but conservatives often regard the evolution of household life, away from the traditional 2.4 family, not as a result of the strain placed on many by conservative social and economic policy, but strangely, as a consequence of state interference. They often regard liberal reforms in areas like welfare, marriage, women's reproductive rights, drug laws and criminal justice as the chief activators of social decline and breakdown. When they move to 'toughen up' laws, or they cut benefits to single mothers, it must be understood that they often do so because they believe, in the long term, the outcomes for those groups and wider society will improve – they think making life more difficult for certain people will result in a sudden magical attitudinal shift. Let's not forget, some of these people in the past genuinely believed you could talk someone out of being gay. It is then you begin to understand why issues like marriage, divorce, abortion and sex education rank highly in the hierarchy of concern for many conservative-minded people – these are regarded as mechanisms by which the institution of family is either enabled or dismantled.

In the United States, prominent Black conservative intellectual Thomas Sowell has argued extensively that the decline of African American communities in the 1970s and 1980s was a direct consequence not of the legacy of slavery or systemic racism, but of generous welfare policies which encouraged dependency and emasculated male role models. In his 2005 essay collection, *Black Rednecks and White Liberals*, Sowell writes:

'The burgeoning of the American welfare state in the second half of the twentieth century and the declining effectiveness of the American criminal justice system at the same time allowed borrowed and counterproductive cultural traits to continue and flourish among those blacks who had not yet moved beyond that culture, thereby prolonging the life of a chaotic, counterproductive, dangerous, and self-destructive subculture in many urban ghettos.'

While this argument applies specifically to the United States, it contains similar themes and ideas which are prevalent among conservative-minded people in the UK. First, the belief that welfare is, when too generously distributed, socially harmful. Second, the belief that punishment is the most effective deterrent against crime and antisocial behaviour. And third, the implicit assumption that certain social groups who interact regularly with both the Welfare State and criminal justice system possess specific characteristics which predispose them to poverty and crime. From this vantage point, poverty and crime are seen as cultural but there is no engagement with the material conditions which produce these cultures. The flipside of this belief is a conviction that these cultural tendencies towards criminality or dependency do not afflict those of higher social castes, when the truth is often that their advantages are, as we have explored in depth, often a direct result of the state intervening to protect or enlarge upon the myriad privileges they enjoy.

While the conclusion that conservatives simply do not care about the poor is not hard to draw, what must be understood is that from the conservative perspective, liberal reforms in areas like criminal justice, welfare and the institution of marriage are regarded as impediments to the development of individual self-reliance – the ultimate conservative aspiration. Conservatives are not opposed to a generous Welfare State necessarily because they are mean but because they often genuinely regard it, in the long term, as a disempowering social force, both for the individual who comes to depend on it, who becomes less likely to reach their potential, and for liberty, generally, as welfare equates to state interference in markets.

What many conservatives misunderstand is how their economic and social policies bring about the very problems which so offend their sensibilities. For example, as divorces increased in the 1980s and 1990s, both Thatcher and John Major threw everything at the wall in a desperate attempt to reverse the tide of failing marriages. Proposals included changes to divorce laws that would create a 'cooling off period' for couples experiencing marital difficulties – a policy that was rightly binned because enforcing it would involve state coercion. The Child Support Agency was established to ensure absent fathers made a financial contribution to the upbringing of their children and included tracking down and pursuing fathers where there was no longer contact. Unsurprisingly, this was also an attempt at a covert cost-cutting exercise – the idea being that money to support lone-parent families would come from absent parents and not the government.

We then got the expansion of the Married Man's Tax Allowance which granted a higher tax-free allowance to married men than single men. But it was in Section 28 – a rule that banned local government from 'promoting' homosexuality – that conservatism's most reactionary tendencies were most evident. Here, more than any other area, the Tories ventured beyond mere promotion of their ideal to stoking bigotry in the face of unconventional family structures.

These were all genuine attempts at keeping families together because, from the conservative perspective, family breakdown is what leads to poverty and deprivation – again, they are partly correct but have it the wrong way round. Lone parenthood itself does not create poorer outcomes for children; factors often associated with being a lone parent, such as high levels of stress and poverty, do. When you liberalise the economy with no industrial strategy to mitigate the inevitable impacts in former industrial communities and then paper over the dereliction with consumerism, cheap credit (debt), off licences and fixed-betting terminals, a rise in family breakdown becomes inevitable.

The twentieth century heralded a golden age of social mobility. The creation of the Welfare State led to record levels of employment, educational attainment and inter-class marriage. But at some point, we forgot that millions had climbed the ladder not simply because of their personal attributes or merit, or indeed because of capitalism, but also because the state intervened to offset the inequalities produced by an increasingly capitalist economy. In relation to the institution of family, there is often a genuine conservative desire to address a social problem but a fundamental misconception of what that social problem is, followed by solutions which are all driven by terrible questions. In this case, 'why are poor families so bad at life?' rather than 'what did we do differently in terms of government policy which might have contributed to a rise in familial dysfunction?'

This pattern is repeated and again, across every area of social policy, creating layer upon layer of overlapping social discord and breakdown. And operating from a conception of the economy as akin to a household budget, many conservatives regard the essential reforms which raised millions up and out of poverty, and into education and employment, as luxuries which must be jettisoned in times of economic turbulence.

In recent years, successive Conservative politicians have brought forward policies and instituted reforms which, like those in welfare, speak to the deeper misconceptions at the heart of their ideology. There was the botched privatisation of probation services which led to worse outcomes and billions in wasted taxpayers' money. Cuts to policing, which led to a rise in crime, which then created demand for 'tougher' laws. A commitment to evict the families of gang members from their homes for the persistent offending of their children – a clear example of throwing petrol on the flames of a problem. The bedroom tax, which led to mass evictions. The scrapping of housing benefit for 18–21-year-olds, which placed young people from poorer backgrounds, likelier to be struggling at home, in education and in the labour market, at increased risk of residential instability and homelessness. Cuts to legal aid, which disproportionately affects the poor who are overpoliced, likelier to enter the criminal justice system and to require legal representation. The rape clause – an exemption from the two-child cap on child benefit for mothers who can prove an additional child was conceived as a result of sexual assault. The scrapping of nurse's bursaries. The Trade Union Act – a wide-ranging crackdown on workers' rights that banned strikes unless 50 per cent of all union members eligible to vote chose them, not 50 per cent of those who voted, as before. Changing the government definition of child poverty and then ditching previous child poverty targets. A public sector pay freeze. Social care cuts. The benefit freeze. Cuts to the NHS. Attempts to hike rents for council tenants. A reduction in mental health beds. And all this while cutting inheritance tax for the wealthy, giving themselves pay rise after pay rise, demanding they should be entitled to second and third jobs and that their legal costs for defending themselves when they are suspected of lobbying are covered by the state.

It is only because Conservative politicians are unaffected by the consequences of their policies (thanks to proximity gaps in both experience and accountability) that they persist with their bombardment of the poor and vulnerable. And an electorally significant section of the British population is as, if not more, remote from this social reality as the Tory politicians they vote for. The affluent, socially conservative strongholds, often referred to as Middle England, despite experiencing less crime, less poverty and less immigration, often hold the most hard-line views on these issues and must be pandered to by politicians of all stripes, one way or another. But this pandering comes far more naturally to top-flight Tories, who broadly share these outdated and harmful views.

It may be hard to believe that Tories go through with all of this because they care. I struggle to accept this myself sometimes. But the conclusion I have reached is that people motivated by a desire to inflict general misery simply do not possess the energy to fuck everything up so precisely. These are the not acts of people hellbent on general social destruction; these are the deeds of individuals whose lack of lived experience in areas like education, the labour market, criminal justice, addiction, homelessness and poverty compels them to tear the complex weft of Britain's social fabric to shreds in extremely careful and specific ways.

The fundamental misconception at the heart of conservativism is that values of family, law and order and self-reliance are universal – that they can be applied everywhere. What they have not yet grasped is that conservatism only works when people have something to conserve. In the absence of the opportunity, money, access to education, legal representation, health services, favourable media representation and ridiculous levels of hereditary privilege, conservativism, as instituted by the UK government since 2010, registers with many not as a benevolent force but as a devilish, toxic, vindictive hostility.

The source of much of this hostility has often been the importing of voracious post-monetarist US business models and practices into the social realm. The role of US insurers in welfare reform has already been discussed but similar misplaced faith in corporate ideology also led to the part privatisation of probation services in England and Wales. The final annual report produced by Dame Glenys Stacey, former chief inspector of probation, published in 2019, found that the housing requirements of offenders were met less frequently (54 per cent of private cases compared with 70 per cent of public cases), there were often inadequate protections for victims and their children when abusers were released and 22 per cent of offenders were liberated from the prison system without knowing where they were going to sleep that night. In both welfare and probation, we see the effects of conservative government's willingness to sell a global crisis in the capitalist system to the British people as a national crisis of public spending.

It is entirely proper that the super-wealthy, the moderately rich and the reasonably affluent are represented politically, and in the Conservative Party they have the best possible ally – a party which is not only run by individuals of considerable economic privilege but also funded to the teeth by them. While the social make-up of the Conservative Party is more diverse today than it has ever been and while it currently draws significant support in post-industrial communities, it remains dominated by those of higher social castes, whose interests are not easily reconciled with those of working people.

There is a world of difference between representing a working-class person's values and advancing their objective economic interests. Indeed, mainstream political parties on both the left and right often rely on vague appeals to poorly defined values in their attempts to adjoin disparate, politically lucrative demographics in broad electoral coalitions, whose economic interests are wickedly difficult to square-off. Thatcher's 1979 campaign slogan was 'Don't just hope for a better life. Vote for one,' but many who did vote for the Tories that year ended up on the dole. New Labour perfected the strategy, based on a belief class was a remnant of the past, with empty but affirming mantras like 'Britain deserves better' and 'New Labour, new life for Britain'. Cameron emphasised the 'Big Society', where we'd all muck in to get things done. And Boris ran exclusively on bringing resolution to Brexit, appealing to people's sense of fatigue on the issue, with the deadly 'Get Brexit Done'. In all cases, the political messaging was designed to entice people in from all social classes without emphasising social class at all. This is to say, these were appeals to values, vaguely defined – aspiration, renewal, community and national identity. The problem with values is that they go out the window when class interests are threatened; the landlord and the tenant may both earnestly believe that everyone has a right to a roof over their head but this principle will be cast aside the moment the rent is overdue.

Values are important. It is right to say that the Tory party can broadly represent the values of voters from all social classes. Values like law and order, family, personal responsibility. It is precisely because the Tories have been so successful in recent years at speaking to cultural concerns such as national identity, sovereignty and free speech (which themselves arise from the insecure economic conditions the Conservative Party has created) that their electoral coalition is changing. The more threatened people feel, the more conservative they become, which makes holding the Tories accountable wickedly difficult, because all a government minister has to do is invoke the trending culture war topic of the day and we go from debating a blackhole of tens-of-billions for a Track and Trace system that didn't work, to arguing about whether men commit crime because Dr Who is female.

The biggest shift in recent years is the number of former Labour voters who are no longer ashamed to support the Tories and having done so once, they become likelier to do it again. But beneath the culture war rhetoric and the scenery-chewing nationalism, the class inequalities which truly constrain the lives of working-class people are widening. This shows that while many working-class people may feel represented by Tory values, in truth, their class interests – wages which rise to meet inflation, a fair labour market, opportunity and social mobility, health and educational equality – have not advanced a great deal. In some cases, they have regressed. Unfortunately, the media they consume rarely reflects the totality of inequality's structural nature.

In the absence of any serious conversation about the economic settlement which has seen working-class wages flatline, a rise in precarious work, exploitative employers and the marginalisation of trade unions, working-class people are presented with a culture war in which to invest themselves. The first rule of political debate that the Tories learnt a long time ago is that the purpose is not to meaningfully convince your adversary of your argument but to win over the unattached audience by any means possible. The perpetual inflaming of the culture war is in truth a fairly simple old-fashioned sleight of hand – don't worry about us hollowing out any chance at broad economic wellbeing, look at the Eurocrats, immigrants and benefit scroungers trying to take your country away. The Tories only aim is to make the Left too toxic in order to win over the relatively politically unengaged floating voters, who cast their ballot on often the most spurious, trivial and emotional impulses. They are aided in this perpetual war by the right-of-centre press, who like to paint the conservative world as a naturally occurring traditional landscape that is beyond question and anything that challenges that world as a dangerous political construction, which makes the task of anyone looking to unseat the Tories incredibly hard. They haven't dominated British politics for the last 200 years because they are especially fit to govern – the Conservative Party's two-century-long death grip on British democracy is a direct result of its powerful backers whose interests it advances without question no matter the cost.

Ordinarily, having set forth such a scathing critique of ideology, I might offer a magnanimous suggestion or three to its more moderate adherents, in the hope that they (perhaps as privately excruciated as the rest of us are publicly by the moral and political direction of their party) may wish to do something about it. I might implore reasonable conservatives to not simply rebrand as 'compassionate', like Cameron did, but to spend more time around the people on the receiving end of the Tory's malevolent and disastrous reforms in welfare and immigration and come to understand that efficiency savings in the short term often mount up as massive cost burdens later on. I might beg them to rediscover their One Nation spirit and recognise that without closer regulation of the market, it simply becomes a driver of poor health, low wages, social decline and political dysfunction which requires greater levels of state intervention to manage. Of course, if they haven't already figured that out by simply observing the utter mess Britain is in, then what chance is there they'll suddenly arrive at the correct conclusion under the audacious guidance of a working-class, left-wing writer, who does not talk, dress or think 'properly'? So, rather than attempt to flog that dead horse, let me simply say this: the Conservative Party has been captured so completely by wealth and privilege that in its present form, it can never represent the economic interests of Britain's working class for one simple reason – to do so, it would have to act against the interests of those it now exists primarily to serve.

17
The Radical Left
Seizing the memes of production

'One cannot expect positive results from an educational or political action program which fails to respect the particular view of the world held by the people. Such a program constitutes cultural invasion, good intentions notwithstanding.'

Paulo Freire, Pedagogy of the Oppressed

The left is by nature better placed than both the right and the centre-left where understanding the working class is concerned. Many working-class left-wingers, whose beliefs are formed as a direct consequence of experiencing social inequality, have natural proximity to challenged communities because they've lived and worked in them most of their lives. They understand the culture, the history and the power relations comprising them and therefore possess a more intuitive sense of what the issues are and how their fellows might come to them. The problem for the left as a political force, it seems to me, is deep confusion at public level about what and who the left represents, the class divide within its own ranks and the ideological wedges these drive between left-wing ideas and working-class communities when the left veers too far from what must always be its central preoccupation – class politics.

The left means many things to many people. In this chapter, I refer explicitly to the radical Left which desires, among other things, a more socialist society. The radical Left of grassroots campaigning, organising and class politics. The Left of public ownership, closer regulation of business and radical downward redistribution of wealth.

The Left of organised dissent, public protest and strategic collective bargaining. The Left that has been consistently willing to challenge power, in all its forms, to advance and secure the many basic rights and dignities we now take for granted.

Whatever your politics, you must surely concede that a functioning democratic society requires a strong left-wing contingent; without it, people are defenceless in the face of exploitation. What the radical Left offers, as well as direct action and other forms of campaigning at grassroots level, is a mechanism by which the centre-left of the Labour Party, the *Guardian* newspaper, NGOs and other moderate left-leaning institutions can be pressured to broaden the terms of political debate.

As discussed previously, many of the advances in areas like workers', housing and civil rights (since integrated into the national mythos as the achievements of politicians) came as a result of campaigns which left those in positions of authority with little choice but to yield. Today, we hear great emphasis on the importance of debate. Of hearing the other side. We are peddled platitudes that disputes can be resolved equitably through civil discussion. Sometimes they can and reasoned debate is often preferable and appropriate. But when it comes to guaranteeing basic rights and protections for working people, the poor and minorities, argument at the public level alone produces little but hot air – organised resistance is what has historically tipped the scales.

The Left exists to antagonise and wrest power from those who possess not only the resources and influence to coerce and exploit working people, as well as the environment and democracy, but also, pivotally, to shape public opinion about the true extent and nature of that exploitation. The expansion of corporate power in recent decades and its tentacle-like grip on mainstream politics and media has led to an erosion of public trust which has exacerbated longstanding economic and democratic imbalance in Britain. Such circumstances provide the Left with a perfect ideological foil, yet, through a combination of a changing media landscape, the absence of a clear, positive, coherent vision of a socialist alternative to the status quo, and an admittedly hostile political and media establishment, the left sees little consistent cut-through in media unless it is being depicted by its detractors.

It is most unfortunate, in an era when conditions are most ripe for radical change, that the Left – a loose association of individuals, campaigns, organisations and institutions which operate independently of each other but towards a common ideological goal – has failed to capitalise on the immense opportunities presented by the proliferation of radical, revolutionary digital technology. Rather than harnessing the power of social media, much of the Left has been subsumed by it. Tribal resentments directed towards the right of the Labour Party, a less-than-inclusive preoccupation with theory and hand-wringing over culture war issues like identity politics and immigration have taken precedence. The chance to leverage social media's radical potential and truly convey to the wider public what the Left does and does not stand for has, thus far, been largely squandered.

When most people refer to the Left today, they invariably mean left-leaning, liberal capitalists, or so-called 'identity politics' (which has a strong liberal, middle-class contingent, often wrongly conflated with the radical Left). It is this confusion, at the level of culture, about what and whom the Left stands for, and what its radicalism truly entails, which has become its primary problem. The case for radical left-wing ideas is hard enough to make, even in simpler times, without these additional folds in discourse occurring in an information ecosystem which thrives on cheap emotional catharsis and hollow spectacle. In this environment, the Left fulfils a role for the other players on the right and in the centre, in politics, and within both traditional and new media. The Left no longer defines itself but is defined by its opponents and detractors. It has been recast as the ultimate cultural bogeyman in a new red scare of algorithmic proportions; it is now, it would seem, both irrelevant and ubiquitous, useless and menacing, full of snowflakes yet rife with warriors. It's whatever the right and the centre need it to be. Rather than a synonym for class politics, 'the Left' is now regarded by millions as being concerned with everything but class. While not every criticism can be dismissed out of hand – many we will now explore – this prevailing perception is, in my view, nothing short of a hideous fiction. A meme that went viral because it felt right and not because it was true.

While radicals often believe this failure to capture the wider public imagination is the result of capital, colluding with media and the

Labour Party, to squeeze it out once again of the mainstream (which is partly true), much like the right, left-wingers are somewhat bogged down by their many burdensome ideological traditions. The problems lie not simply in the lofty objective of instituting a more socialist society (a perfectly reasonable if slightly ambitious aim) but in how leftists often carry themselves in the public arena as a result of their under-exposure to working-class communities and overexposure to radical socialist subcultures.

Before you can truly become a left-wing radical, you must first accept and then internalise certain beliefs about yourself and the world. This political osmosis invariably occurs in youth but, unlike other youthful traits, these left-wing precepts are not always moderated by age or subsequent life experience. You must forsake the language of personal responsibility completely – everything wrong with the world is the fault of a system, structure, elite. You must regard the material world and the development of history as a convergence of escalating, inter-locking crises, attributable to one causal factor – capitalism – and not as an immensely improvised mess, endlessly rebuilt on top of itself, arising from the general chaos of human endeavour, directed by over-confident elites operating with imperfect knowledge. And in the face of this ubiquitous menace, you must speak always in the language of perma-crisis in your attempts to mobilise the masses – naively over-looking how the average person, with limited resources, becomes likelier to develop and then act on conservative impulses when their security instincts are threatened.

All of these traits, which are the attitudinal and behavioural expres-sions of a developing adherence to left-wing ideology, are then maintained by an overbearing ideological superego, which polices your thought and conduct much like a deity might to those of varying religious conviction. And in the face of public scepticism, indifference or hostility, whether justified or as a result of biased media coverage, the leftist acquires a particular emotional attitude which fortifies their core beliefs against attack – even when those beliefs are false or require refinement. This uncompromising fixity is no competition for a conservative's ability to change course at a moment's notice or a centrist's brass neck to do the same but make it appear less of a compromise. The radical Left is like an ideological battle-tank that rolls ambitiously into the public square but then struggles to man-oeuvre within the confines in which it has willingly placed itself.

Central to the Left's collective identity is the notion of struggle which arises from its reading of history as a war between classes with opposing economic interests. But too strict an adherence to this belief renders almost all forms of compromise unthinkable, which places the Left always at a strategic disadvantage politically: while the right and centre are able to move left and right, triangulating as they go, depending on the electoral or media requirements of the day, the Left is bound by the strictures (and vanity) of the ideological overseer. Unable to confront this, many on the left simply double-down, descending deeper into philosophical or theoretical rabbit holes where they can once again derive a sense of meaning and purpose, a sense of victory in defeat, and where the tricky business of confronting polit-ical reality can be indefinitely postponed.

Ultimately, the true radical's aim, subconsciously, is not simply to create a socialist republic but to be validated by history having done so in a very particular manner. The Left's struggle is not simply for a fairer society but to emerge as the victor in the long-standing battle with its nemesis, capitalism. In this context, compromise is akin to admitting you were wrong – something many radicals are constitu-tionally incapable of doing.

These attitudes and beliefs become assets and liabilities in equal measure; the deep sense of surety they engender propel the leftist for-ward into action, helping to establish them within a wider peer group as well as more easily identifying and organising against power struc-tures. But those beliefs – when too strictly adhered to or when emerging uncritically from group-think – may also create an unb-reachable political glass ceiling, locally or otherwise, where the compromises so often necessary to advance an agenda, appeal more broadly to a community or resolve an internal dispute, are regarded scornfully by small but persistent elements as collusion with [insert an enemy of your choice]. It is here where the tedious factionalism on the left begins – one side decides the other has broken ideological rank and splinters off from it, before later consuming itself. The Left's successes and failures originate from these core beliefs and their accompanying attitudes and behaviours. What grants it resilience and focus also renders it incoherent and impotent. What allows it to plant deep roots locally, leaves it rudderless at scale.

The radical construes the world through a collectivist lens. This mindset produces a culture where individuals are encouraged always to look externally for all of their answers, when occasionally the harder truths lie closer to home, often on reflective surfaces. Left-wingers are often very certain of the motives of those they would assail, whom they think self-interested, myopic and disingenuous, while exhibiting a self-defeating lack of concern for their own obvious blind spots, prejudices and egotistical impulses. This differs from the conservative notion of imperfection, where the individual is assumed, on some level, to be corruptible and self-interested – a far more realistic appraisal of the average person if not quite as romantic as the left's 'salt of the Earth' notion of the working class and themselves.

Because its eyes are trained outward, the Left has trouble seeing itself the way others, who are not enmeshed in the culture of the left, perceive them. The dearth of willingness to self-reflect (a neoliberal indulgence which distracts from the class war) means political problems are often as misdiagnosed as those found in the personal lives of many on the Left, and defects of thought and strategy are (like the hard truths that keep them tossing and turning at night) swept under the carpet or sarcastically dismissed. Criticisms of the Left, like every other problem, are viewed as merely capitalist mirages – bad-faith arguments insincerely held or rooted in mainstream media thought-conditioning. Whatever your opinion, the left-winger will have read a book which easily refutes it. What the leftist often hasn't read are the many books which many believe refute popular leftist claims and positions, and the historical interpretations they often hinge on – they are capitalist propaganda, you see.

Corbyn's first appearance at Prime Minister's Questions as leader of the opposition in 2015, despite the high level of anticipation to see him square off against a front bench of Tory toffs begging for a dressing-down, was so deeply anticlimactic precisely because radical lefties, in this mould, seem to relish the eschewing of basic presentational conventions. What is a close-skin shave or a fetching fitted suit, or any other minor presentational compromise which might broaden the Left's superficial appeal, to a well versed radical with history on their side? This is not to suggest that had Corbyn scrubbed up a little better on television, that Rupert Murdoch and the Blairites would have warmed to him – that would be a ridiculous claim to make – but there's something to be said for understanding where the proverbial bullets are going to come from when you enter the public arena, should a radical somehow find themselves in the unlikely position of leading the Labour Party. The clear hostility Corbyn faced from day one as leader of the opposition was compounded by his lacklustre media performances and this ill-considered presentation style signalled the more fundamental problem – his team had no strategy. He often gave the impression that he was Labour leader because he had lost a bet. Corbyn is far better suited to grassroots campaigning, where he works a crowd like few politicians (or activists for that matter) can. But the idea that he possessed the skillset to take the Labour Party back into Downing Street was wishful thinking – not because he was radical, but because his specific talents were not suited to the current context.

Ironically, despite this outward impression many left-wingers give of relishing caring little for appearances and being all about the substance, most left-wingers are, in truth, far too preoccupied by how they appear to their comrades. They often overlook the significance of presentation, small modifications to which may render their many legitimate criticisms of capitalism and Western foreign policy less jarring to those outside their group.

The Left splits along two lines – the doers and the talkers. This is not to suggest that discussion and debate are superfluous in the struggle for class equality, but simply that those (like myself, lately) whose contribution to the cause can be measured mainly in airtime and column inches, tend to be less involved in the day-to-day business of organising. It should, therefore, come as no surprise that even on the left, class dynamics and imbalances are rife. This is as true in the arts and media as it is in politics – middle-class people take centre stage at public level and those of lower social castes are left to do the heavy labouring.

For a certain left-winger, whose presence is felt primarily online, there is little real-world engagement with issues or communities, only the endless commentary, analysis and occasionally humorous, though often impenetrable, salvos of political and historical marginalia. It is right that every movement is underpinned by a rich intellectual tradition, but on the left this often takes centre stage – driving an additional wedge where it need not be.

I recall a conversation with leading anti-poll tax activists, who organised Glasgow's poorest communities to resist the unfair levy. These were working-class people with few qualifications, whose street credibility and willingness to sacrifice eventually galvanised a nation, resulting in the tax being scrapped. They joked, as they reminisced about the days of holding public meetings in the stairways of dilapidated tenements, that the intellectuals would occasionally turn up, whip out their literature and read – in an attempt to superimpose leftist ideology. This, I was told, often elicited rolls of the eyes and under-the-breath laughter from local people who smelt the opportunism a mile away. In truth, the anti-poll tax movement succeeded (like all campaigns) due to the action taken by working-class people who never made it to university – not the bookish confidence of sarcastic academics.

In the absence of any genuine political power (and the responsibility which always comes with it), leftism becomes an intellectual plaything for some, and certain left-wingers grow very sure of themselves and their theories as a result of their grand hypotheses having never been tested. It's a strange kind of moral superiority that's only possible when you've never been responsible for anything. They come to believe that at every crucial juncture in recent history, crises may have been averted had only the Left been in charge. From foreign policy to economics, the Left enjoys the luxury of playing casual observer – no truly radical, left-wing leader has been proximate enough to power to hold sway over a decision of great national consequence in the UK – and while this does not invalidate the Left's criticisms of the Western consensus, it does engender a lack of humility, a moral surety and an over-reliance on counter-factual arguments which are, by definition, hard to verify.

Many leftists then develop a false sense of confidence in their worldview. One which may find expression in an unjustifiably strident or patronising attitude. They begin to lose touch with the notion that much of their outlook is theoretical. This behaviour endears them to others in the community, allowing them to carve out niches and delight in how difficult or cynical they are. How obstinately they behave. And how obscurely they can construe current events and history. This certainty in themselves means they rarely concede or apologise. If they do modify their view on a topic ever so slightly, the new position is adopted covertly, to preserve the veneer of intellectual continuity. And in this emotionally stilted, dishonest approach, they model a behavioural template to the generation coming up behind, who turn to them naively for a wonderful example of what a left-winger looks like.

Left-wingers are often keen to portray an image of intellectual and moral independence but often they seek cues from those around them, not simply about what or how to think but how to engage in debate and even what to say more generally. This is most apparent in the age of 'trending' topics, where the group-think and utter lack of originality are more visible than ever before.

Taken together, the manifestation of ideology acquired in youth and reinforced over subsequent years renders the left highly effective in a local setting (where working-class radicals make their values visible within a community) but arguably less effective at scale – where middle-class pundits appear remote, highly unaffable, smug and idealistic. The Labour Party manifesto in 2019, which many claimed was popular, was a perfect example of leading with the chin. The document simply contained too many radical policies which, individually, may have sounded good (and polled well in a brief phone call) but taken together, in the current cultural context, were quite simply too good to be true. That is not to say a radical prospectus is not possible, but presenting one requires an understanding of the environment it's being launched into. People were not ready to hear such a radical message after a decade of Tory austerity in which they were bombarded by negativity. Curiosity about solutions beyond the accepted parameters of debate must be cultivated and nurtured, not regarded as self-evident. People don't vote based on individual policies; they vote on a leader with a strong vision that promises change nested within continuity.

The insistence from some on the left that it won the arguments but lost the election is simply further evidence of a stubborn refusal to concede even the simplest, basic truth. It is of course true that the press was extremely hostile, but this was entirely predictable. Fighting two general elections without a coherent media strategy, Corbyn often taking bizarre leaves of absence from the spotlight for days, creating unnecessary speculation, was absolutely central to the level of confusion publicly about Labour's position on Brexit. His passive media style was an open goal to anyone with a desire to smear him, with the intention of derailing his political project.

Whether campaigning, media punditry or in the realm of traditional party politics, in the age of social media, the left can't simply pick up the digital tools to convey its message more effectively, it has to sharpen its self-awareness with regards to how radicals and left-wing ideas are now viewed by the general public and develop more effective media strategies to deal with that. There has been some promising progress, on both sides of the Atlantic, though, ironically, working-class voices remain absent – no doubt a consequence of intellectual gate-keeping and self-defeating academic snobbery.

It is in the media space (old and new) that many left-wingers, through a lack of self-awareness and an over-immersion in left-wing cultures, histories and theories, can disadvantage themselves and their causes. Corbyn had no real strategy to deal with the powerful forces his unlikely rise to leader of the opposition inevitably unleashed. Indeed, such a strategy would be perceived as 'spin', which is to demonstrate a profound unseriousness in an age defined more than anything by first impressions.

This is quite typical of the nonchalance on the Left, where matters of public relations are concerned. Public relations are now synonymous with New Labour and so, in a sense, it would be somewhat shameful on the Left to be seen obsessing over narratives and optics. However, while Blairism did become a synonym for 'spin', nobody was in any doubt as to New Labour's agenda and, more to the point, who was in charge. New Labour acted with intention and discipline and, while I disagree strongly with the agenda these skills were deployed in service of, it cannot be denied how effectively it operated in a fast-paced, rapidly changing media environment. That clarity of focus, in service of a more radical agenda than New Labour ever attempted, supplemented by the radical Left's undeniably effective ground-game, is the only route to political power, either within the Labour Party or as a distinct new entity.

In a culture which is now driven almost exclusively by various forms of media, left-wingers will need more than podcasts and YouTube channels to drive home their message, and whether the left of the Labour Party insists on its battle for the soul of that institution or whether it breaks off to create a new political party, its success will lie in part in its media strategy, not simply in how it pertains to the public but also the wealthy special interests which are sure to attack it. Blair got in bed with Murdoch. It was distasteful but it was a strategy nonetheless. There is no road through the British media landscape without compromise of some form and no route to political power without an effective media strategy – though what that looks like for a genuinely socialist party, in the unlikely event one does arise, remains to be seen.

Social media provides new opportunities to reframe the issues of the day from a left perspective but has thus far acted mainly as a broadcaster of the narcissism of small difference coursing through left-wing online discourse. Social media is utilised too often as a sneer dispensary, where people on the left talk among themselves about the inner workings of the Labour Party and their long-running battles with moderates, or, increasingly, impulsively crisis-hop from one calamity to the next, in an attempt to weave together a coherent anti-capitalist narrative. While there is value and legitimacy in these pursuits from a distance, they simply cement the impression left-wingers are inward-looking, ideological obsessives, because the aim here is not to speak broadly to the wider public but usually to mobilise itself.

The Left has two means of leaving a favourable impression on the public mind: by making its values visible on the ground in communities, or through traditional or new media. With respect to the former, the Left's ground-game remains strong. Indeed, many campaigns and collectives have responded extremely well to the threat of far-right populism by quietly entering communities, not to moralise, but to address many of the legitimate core grievances which give rise to it.

In recent years, we have seen a rise in tenants' unions across the country as well a spate of wins for workers in disputes over wages and conditions in both the private and public sectors. These local and national victories do not occur in a void. They are not inevitable. They are the result of highly skilled, dedicated campaigners working in partnership with communities. This work succeeds because its aims are tangible and linked always to the class concerns of local people. And while collective bargaining is distinctly of the left, workers do not need to pass ideological purity tests to join a trade or tenants' union.

Too often, the activists likelier to generate more favourable perceptions of the Left at the public level (because they are working class) are those who are involved in the grind of campaigning. This is particularly true of female organisers and campaigners. They are rarely glimpsed at the wider public level. Their lives are dominated by the causes they dedicate themselves to and demands on their time and headspace are compounded by the fact they – by virtue of social class, gender and often ethnicity – may also be experiencing myriad forms of poverty and deprivation themselves. These activists do not produce 'content'. They do not regularly dispense their views from a distance on the pages of national newspapers. They do not view meetings as opportunities to roleplay as philosophers. And though they may desire to take more visible roles, they are often plagued with self-doubt.

The media spaces, new and traditional, are then dominated by academics, intellectuals, commentators and pundits who often lack the common touch and the proximity of real action which would endear them to working-class people. They often (though not always) speak with affectations which place them firmly in a middle-class arena – like I do in much of this book. They express ideas in accordance with how they have been taught in university or indeed other left-wingers they've been influenced by and too often demonstrate either a lack of ability or zero willingness to modify this. This problem is compounded by the fact the activists on the ground – particularly the men – often look to these more socially remote, intellectual figures, locally and globally, for ideological instruction – which they may trust more than their own vastly superior political instincts.

This quirk of left-wing culture was perhaps most pronounced in the mid-2010s as the lines between left, right and centre began to blur with the rise of nationalist populism and so-called identity politics – a constellation of social movements which focus less on material conditions and class and more on the subjective experiences of minorities and marginalised groups. While the Left, in more ordinary times, may have found natural allies in either camp (EU scepticism as well as equality for minority groups are part of the radical tradition), leftists instead became mired in the ensuing culture wars. Now working-class people would be split into deserving and undeserving categories by many leftists, on the basis of their views on immigration or intersectionality (the intellectual framework of so-called identity politics where aspects of one's identity, such as gender, ethnicity and sexuality, are seen as compounding factors in their oppression). Others, understandably, backed away from these debates entirely, with the rise of Corbyn providing a brief period of illusory unity.

Certainly (from my point of view at the time) there was a shortage of activists willing to push back and assert basic truths about mass immigration as doing so brought into tension two important principles on the left – class solidarity and anti-racism. But not all immigration concern is rooted in racism or xenophobia. Many understood this but too few were prepared to mount a response to a gathering storm of middle-class liberal hysteria and aggressive New Right opportunism.

A vacuum opened up where the traditional class politics of the left may have been advanced, into which demagogues like Farage and thugs like Tommy Robinson, in collusion with global players like Steve Bannon, wisely manoeuvred. It was they who then adopted the language of class solidarity and, while offering no real economic solutions, effectively endeared themselves to working people the country over, while community leaders and thinkers on the left, to whom many on the ground looked for instruction, backed away from these controversial topics or chose the path of least resistance.

The myth then began to pervade the Left that class equality need not be its sole purpose. That it could accommodate and integrate identity politics. That these were two branches of the same tree. But the problem with identity politics was never that it sought to advance the interests of the marginalised but rather that it came with a style of remote online activism, devised on elite university campuses in the

United States, which, as well as utterly devoid of strategy, was tone-deaf to social class dynamics.

Let's briefly examine a central tenet of identity politics – the notion of privilege. Privilege is in many ways the barrier to proximity. Many in working-class communities take the term privilege literally and, often, this word is deployed pejoratively and functions as a slur. In communities where there has historically been an absence of opportunity and an overexposure to violence, economic hardship and terrible governance, where locales are in visible decline and media coverage is always grim, privilege is seen as existing 'over there'. With even a modicum of understanding of class, it would seem obvious that middle-class activists accusing working-class people of possessing privilege was always going to end terribly. The vast culture gap between working-class people coming to this new language cold and the middle-class students and activists who had been immersed in it for some time (themselves battle hardened by the culture war which had, until the mid-2010s, been confined to online subcultures and largely out of public view) meant discord was inevitable. When you become so immersed in a theory and have little contact with the wider population, you come to perceive natural scepticism and disagreement as the enemy – not an opportunity to refine ideas and develop new strategies.

Many on the left understood the complexities at play as the culture war began to take shape in the mainstream but informal social pressure within left-wing communities inhibited many from taking the difficult positions necessary to truly move through these contentious issues with confidence. As terms like 'trigger warnings' (the common courtesy of forewarning people prior to the discussion of sensitive topics which, while kind and compassionate, has no evidential basis in addressing trauma), 'safe spaces' (which have always existed but were not referred to as such) and clumsy, tedious refrains like 'check your privilege', erupted through social media, perfectly legitimate criticisms, and questions regarding them, met with hostile, closed responses.

The fact so many were surprised that working-class people who live in conditions of economic humiliation would resist the notion they are privileged is evidence of how socially remote many of these activists were. For a branch of post-structural thought which rightly asks us to consider the meaning and impact of words on others, as well as their historical context, you'd think someone would have twigged that the word 'privilege' is loaded with centuries of connotation – most working-class people associate it with wealth and use it as an insult.

But these new ideas and concepts were not up for discussion. To mount any criticism whatsoever, whether about the changing nature of language, the shifting parameters of debate or the countless new terms doing the rounds, was to mark yourself out as suspect. Much of what occurred in the mid-2010s was not radical in the traditional sense – it did not challenge power structures – but was simply a new strain of middle-class domination for the social media age: a cultural sensibility which relied on passive aggression, language and speech policing, punitive shame cultures and smears dispensed from a safe distance to achieve a cooling effect – people began to keep their views to themselves rather than be persuaded to adopt new ones – which created the illusion of change.

The failure of some on the left to enter the arena and contend with nationalist populism and identity politics at the moment they became simultaneously mainstream handed a blank cheque to right-wing agitators who have since successfully recast themselves as speaking for the common man and woman – and recast the left as a distinctly middle-class project.

The Left's central preoccupation must always be class politics. Class politics which is fully inclusive of all protected and minority groups, of course, but which organises around tangible class-related grievances and objectives. Class creates the connective tissue sufficient to build a mass movement carrying a message of the necessary weight and depth to move society's economic needle. Whether racial minorities, the LGBT community, people living with disabilities, addiction issues or mental illness, social position remains the decisive factor shaping the experiences of most who count themselves among protected or minority groups. Women are more likely to experience gender-based violence when they live in poverty. LGBT people are more likely to experience poverty because of discrimination in education and the labour market. Young people from BAME backgrounds are likelier to be stopped and searched when they live in poorer areas. The arbitrary characteristics of sex, gender, skin colour, sexuality and so on are, of course, important but it is the compounding factor of class which invariably leaves women and those from minority groups more vulnerable. Social class offers a fuller explanation of the material conditions of poverty, exclusion and discrimination in which many from protected or minority backgrounds often find themselves. Asserting this is not to dismiss the role that cultural identities play but is simply to recognise that the compromises often made by institutions in the face of pressure rooted in cultural grievances as opposed to economic ones are often cosmetic or rhetorical in nature, offering little with respect to widening opportunity or raising living standards.

People from all backgrounds should be welcomed to join the Left in this project. Every effort should be made to ensure that all unions, campaigns and actions are inclusive and democratic and that discrimination of all kinds is dealt with. Better yet, steps should be taken to build on the last few years of progress with respect to women and minorities assuming more leadership and public-facing roles, placing them at the heart of class politics, which has historically been dominated by men. That being said, the Left cannot be all things to all people and the aspects of identity politics which are indulgent or toxic (some of which I have outlined briefly) should be compassionately discouraged. The Left must be never become hostile or dismissive of identity-related issues, nor adopt a rigid, reductive view of what a working-class person looks and sounds like, but the radical should never be afraid to emphasise what the Left stands for – all-inclusive class equality – and to deploy strategies and tactics in service of this aim without fear or favour.

The Left is so preoccupied with the urgent task of radically transforming society that it has neglected to consider the possibility of radically reforming itself. Modernising voices must come to the fore, speaking not simply of capitalism, current events, theory or history but of coherent strategies to re-establish proximity and clear lines of communication (both ways) with working-class people. Strategies which are rooted in a painfully frank assessment of where Corbynism failed – not everything can be pinned on the Murdoch press. Working-class radicals must be more confident in themselves and push into the public realm (and past the neck-bearded chin-strokers), with communication strategies as detailed as any other form of organising. They should relish the challenge of becoming a public face of the Left and resist chatter from the margins that engaging with media is ideologically impure. This is not about creating a false impression or 'spinning', it's about communicating a message with discipline, intent and style, which bring left-wing ideas and concepts directly to working-class communities, rather than leaving them on the table to be mediated by bad-faith actors mining for outrage clicks.

Developing a sensible language around the theme of personal responsibility, which recognises the role the individual can play in bettering their circumstances within an unjust system, and accepting that promoting wholesale systemic change as the only route to a better life is not very empowering might be also be worth considering. Advocating personal responsibility is only a problem when people are commanded from on high to pull themselves up by their bootstraps by those who have never tasted desperation. Recognising individual agency within the context of systemic inequality should not be taboo or controversial on the left and that many still resist this notion on ideological grounds represents an unnecessary act of political self-harm. The key to personal responsibility is that when we encourage people to take it, we understand intimately what we are asking them to take responsibility for, and we do so with love, compassion and understanding – not to shame, humiliate or threaten.

Formulating a critique of capitalism which recognises it's all most people have ever known and that its emergence centuries ago emancipated many from a previously feudal system might make the bitter pill that eventually it has to collapse like every other economic system a little easier to swallow. Conceding that the most immense period of human progress has run concurrent to free markets is not giving capitalism's terminal defects and contradictions a free pass. A serious critique involves acknowledging the myriad advantages many people associate with free markets – competition, efficiency, innovation, choice, liberty – and using them to pose an alternative which reassures those anxious that much of what they enjoy would remain, in some form, but how wealth is distributed and how institutions like corporations are structured and owned would change. As bad as things get, talk of radical change without a clear definition of what shape an alternative might take is scary. Just as the Left is very comfortable with being morally right but not so comfortable with doing what needs to be done to achieve power, the Left is also too adept at describing the problem while skimming on the detail around solutions – don't I know it. Perhaps more than anything else, a balance must be struck between rousing rhetoric which emphasises crisis and decline, and a hopeful, positive vision of the future. When your entire political project hinges on the present order collapsing, thus proving your hypothesis, you are always at risk of being seen as an accomplice when the world starts falling apart. How does the left reconcile the need for a new kind of society, and the necessity of pointing out the endless problems in the one we have, with the grudging recognition that most people feel it's still better than most? Does the Left possess the humility to admit such a thing while still remaining true to its principles? Or would such a basic accommodation represent one compromise too far?

Given we are never short of radical ideas on the left (where everything and everyone else is concerned), perhaps the average left-winger might dare to reflect, in the privacy of their own mind of course: when was the last time the Left truly tried something new?

18

Populism

Populism is a misused term. It has become a synonym of sorts for the far-right in recent years, in part due to the odiousness of its many agitators and ringleaders around the world who have successfully stoked nativist sentiments in an age of galling class inequality. While often referred to in global terms, populism is context-specific – the populism of Donald Trump would likely not fare well in Hungary, just as the populism of Hungary's Viktor Orbán would be rejected in the UK. However, it was in Britain that one of the most consequential populist revolts was staged, in 2016, when a slender majority of the UK electorate voted to leave the European Union.

Populism is as old as democracy itself. It swells in predictable surges when conditions allow it and since the turn of the millennium, a convergence of global crises, economic downturns and the rise of social media has hastened its emergence. The attacks on 11 September 2001, the subsequent War on Terror, followed by a financial crash, over a decade of austerity as well as a refugee crisis in 2015, worked together to create a growing sense of unending calamity. Of a world coming apart. The strain on working-class incomes, the rise of precarious work and growing culture war alarmism led to higher levels of stress and fear. And as the electoral coalitions which were previously key to winning power began to fragment, political parties grew weaker, regressing into factionalism, thus forcing leaders onto ever more uncomfortable terrain in order to remain electorally viable.

While populism is often attributed to one main cause, the revolts we have seen in recent years, particularly from the right, arose from this combination of global, cultural and institutional crises. But whatever the circumstances, populism of all stripes contains two primary ingredients which define it: it must claim to speak on behalf of ordinary people and these ordinary people must stand in opposition to an elite establishment which actively works against their interests.

Left- and right-wing populisms align in their deployment of these principles but differ in how elites and ordinary people are defined. The populism of Corbyn's 'for the many, not the few' and the Occupy movement, for example, coalesced around more objective socio-economic concerns related to wealth accumulation and downward distribution. Right-wing populism, in contrast, focuses more on socio-cultural grievances which arise from issues like immigration and national sovereignty, nested within a broader patriotism, which act together to stoke a volatile nativist sentiment.

While populism has been cited as the cause of a string of political upsets in the West, from Brexit to the rise of Trump, the prevailing notion that grievances arising from globalisation are driven solely by lies and conspiracies (or that every course of action proposed by a populist is simply a manipulative attempt to offer easy solutions to complex problems) is dangerously wide of the mark. For at the eye of a populist storm often lurks an uncomfortable truth which draws many into the populist vortex – that the global system is rigged.

One's willingness to accept this fact is often directly relative to how generously they believe they've been cut into the action. The more well-to-do they are, the more tolerant of the status quo people tend to be, and the more averse they are to talk of radical change. From this more affluent, metropolitan vantage point, problems are always seen as burdensomely complex and calls for radical action are resisted and regarded as knee-jerk. While many problems are indeed wickedly complicated, and knee-jerk reactions are undesirable for obvious reasons, what is not complex in the slightest is the basic principle which seems to apply the world over – there's one rule for the wealthy and another for everyone else. It's the failure of some to call that out that leads to a build-up of frustration which inevitably seeks an outlet. Being subsequently labelled a populist for having the balls to call the system what it is – fundamentally unjust – simply adds insult to injury and hands populist agitators a political open goal.

The circumstances preceding the Brexit referendum have deep social, cultural, economic and political roots. Indeed, many of the areas we have covered in Act One, pertaining to matters of class, oppositional interests and culture gaps around poverty and language, converged to produce the perfect populist storm. While the electoral coalition which delivered Brexit was not distinctly working class (it relied heavily on more affluent British nationalists whose proximity to poverty or indeed to immigration is debatable), it is clear that communities which had previously suffered due to deindustrialisation were key and acted as a battering rams, granting a Leave campaign funded to the teeth by wealthy special interests a veneer of working-class authenticity which eluded Remain – for obvious reasons. While salient issues like immigration became focal points, what must be understood are the ulterior dynamics which ripened so many working-class voters – an overwhelming majority of which voted Leave – for political mobilisation on such a scale.

Without question, the decisive political issue which led to the surge in populism in recent years was the swell in immigration concern. While this had been simmering since as far back as 2005, it steadily increased throughout the course of the decade as part of a broader narrative about increasing political and economic integration with Europe, remote bureaucracies, unaccountable power centres and elite mismanagement. While in the lead-up to the referendum there was much focus on the exploits of populist self-publicists like Nigel Farage or racists knuckle-draggers like Tommy Robinson and Katie Hopkins, figures such as these can only rise to prominence when the masses begin to lose faith in a nation's institutions.

Botched foreign wars fomented distrust. Anti-elite sentiment was then further stoked by a financial crash caused primarily by bankers whose collective crimes went largely unpunished. The failure of the political class to speak truthfully to the causes of the crash and to then hold those responsible to account was perhaps the gravest misjudgement of all. One arising from the uncomfortable but irrefutable fact that British political leaders had deferred to bankers in the development of policy to deregulate the financial sector, blissfully unaware of how entangled global financial institutions had become.

As then-Prime Minister Gordon Brown toured the world at the height of the meltdown, keen to be seen persuading world leaders of the urgent need for radical action to underwrite the global economy, the situation was framed by him and much of the media as nations 'fighting back against the recession'. Their repeated failure to adequately address the elephant in the room, which was, in Gordon Brown's own words years later, 'banks running capitalism without capital', led to a sense of legitimate anger and understandable resignation in many communities that the system was rotting to the core. As people lost their jobs, unemployment rose and the stress levels were dialled up by hyperventilating media coverage, the wealthiest always seemed to experience little-to-no consequences for their misdeeds, while the working classes were thrown into economic uncertainty.

It's in this context that people began to develop a more pathological scepticism of politics and media, and rightly so. And those whose prior experiences of authority and public institutions were not exactly positive naturally found this an easier leap to make. This is where the gulf between those whose assets and interests were protected by the bail-out (the upper and middle classes) and those who were cast into the economic wilderness so that it could all be paid for (the working classes and the poor) really began to widen. As the disconnect between British institutions and great swathes of the country grew and the gulf between social classes widened economically and culturally, traditional media's grip on the UK's fractious but largely agreed sociocultural reality began to loosen, as social media platforms acting according to their own perverse incentives made their fateful and ominous incursion into our lives.

A spate of seemingly tangential and unrelated scandals – MPs expenses, cash-for-access, the mis-selling of PPI by the same institutions bailed out by government, Jimmy Savile's serial abuse of children at the BBC, the findings of the Chilcot inquiry – created a new narrative in many minds. One which often ran contrary to the intuitions and pronouncements of public officials and the press. And as the pain of austerity was increasingly felt across the country, a new paranoid mean-spiritedness crept in. The rising sense of anger, stoked by sections of the media whose interests in sowing the seeds of a populist revolt were clear, heightened already significant levels of individual, household and community stress. It was this emotional volatility within sections of the public which incentivised EU-sceptics in politics and media to transpose the decline of British institutions onto the European Union – fears about the pace and scale of immigration being the vehicle by which millions were eventually mobilised.

Taking Back Control – the Leave campaign's marketing masterstroke in the referendum debate – while deployed ostensibly in reference to Brexit, actually spoke to every other area of people's lives, where they quite simply felt the world was moving on without them, politically, economically and culturally – because it was. The Leave campaign exploited this sense of social isolation, economic immiseration and cultural alienation to devastating effect, successfully using the topic of immigration as a Trojan horse to roll-over the opposition – a killer strategy which neither the left nor centre possessed the political or cultural proximity to adequately deal with.

Populists appear to the common man and woman as straight-shooting speakers of hard truths. This impression is created not because they themselves are particularly honest but as a consequence of the highly moderated, media-generated discourse through which they emerge, initially as ratings-winning freak-shows, mavericks and underdogs. In this tepid media environment, where more radical voices are conspicuously absent, which is marked by a collective inability of those of a more middle-class metropolitan persuasion to truly grapple with the structural nature of inequality, the populist's presence becomes all the more transfixing. It's often the first time some people have heard a media personality speak directly and consistently to their marginalised concerns or experiences.

Once endeared to 'real people' by simply stating a basic fact, such as 'elites run society in their own interests', or by cynically conflating Britain's structural inequality with mass migration – for example, on a national panel show where those with opposing views do not possess the frame of reference to adequately deal with populist falsehoods – the 'real people' (working-class folk who have been poorly treated and represented their whole lives) then become susceptible to right-wing populism's more extreme traits. While many cite Nigel Farage as the ultimate architect of Brexit, the referendum result arose from the chaos produced by variety of political figures acting in their own political interests at different points, which speaks to the less noted political roots of the crisis.

First-past-the-post produces duopolies which by their very nature cannot be truly representative. As Britain grows more divided, the two main parties must pitch themselves broadly in ways which create problems for them later on. As they contort themselves to triangulate lucrative new demographics while attempting to simultaneously retain what they regard as their traditional bases, they become politically overstretched and ideologically redundant. Meanwhile, we have a House of Lords still rooted in the ridiculous principle of hereditary privilege, and those who find themselves there not as a result of birthright are often placed there for political expediency rather than because of any inherent ability they may possess. In essence, Britain's main democratic institutions are misaligned with swathes of the population; people want their lives to improve but parties who desire the power to do that become constrained by the winner-takes-all imperative of Britain's electoral system.

Rather than setting forth a clear, honest prospectus, outlining a vision of Britain which is attractive to voters, parties are forced to build unsustainable electoral coalitions by underselling the longer-term implications of their proposals until after they get elected. Once in power, the expectation management begins as the reality of governing sets in and the seeds of their political demise are sown. While populism is often defined as an attempt to gain popularity by providing easy answers to complex problems, the electoral orthodoxy which arises from Britain's first-past-the-post system necessarily demands that even mainstream parties must innovate ever more perverse strategies which themselves hinge on half-truths and falsehoods – presenting the UK's problems as simpler to solve than they are – while promising solutions which in the long run cannot be delivered in the manner specified or, as has often been the case, simply won't cut it.

The notion that Britain is a meritocracy is as big a lie as any the Leave campaign peddled. The belief in trickledown economics was as groundless as the notion the flatlining wages of British workers was a result of migrant labour. And the biggest mistruth of all, that the class war was over, was as populist an assertion as Boris's 'oven-ready' assurances over the Brexit negotiations. In this vapid political tussle between two hollowed-out political edifices, whose paper-thin promises to change Britain have led to more of the same for decades, even the most talentless, self-interested populists begin to look, sound and feel authentic – the final nail in the coffin of slick, 'phony' politicians.

Populism as it is currently defined (from the centre ground) is seen as merely a means of deliberately stoking division. While that is certainly true, it is always foreshadowed by the failure of mainstream parties to address fundamental inequities in Britain's economic structure. In essence, while mainstream parties are not regarded as populist because, rhetorically, they emphasise themes like unity, the policies they undertake, by their nature, create division in the longer term because they inevitably widen the gulf between the winners and the losers. The label 'populist' in the more recent context was increasingly deployed as a pejorative, reflexively placing out of bounds any policy prospectus which might disrupt the privileges to which the middle and upper classes have become accustomed. Any disturbance to the status quo is politically intolerable to its beneficiaries – not because they do not want a fairer society, necessarily, but because they are so craven in the face of the powerful interests whose dominance hinges on structural inequality that they cannot bring themselves to even mouth the word 'class'.

Higher taxes on billionaires – populist. Economic nationalism – populist. Rent controls – populist. Universal Basic Income – populist. Free integrated public transport – populist. Closer regulation of corporations – populist. Political realism (like merit) is thus defined according to the aspirations, values and preferences of those who have benefited from the current settlement, whose sensibilities are not offended by it and whose gains are protected by it – anything outside this is populist.

While I appreciate many of you are exhausted by the topic of Brexit, and many more have accepted it and moved on, it is clear from the aftermath that the lessons have not been learned as to how class dynamics produced this tumultuous political event. It is too great a simplification to pin Brexit on Dominic Cummings, or dark money, or Steve Bannon, or right-wing tabloids – the result was not inevitable and could have easily gone the other way. It was the failure of political figures, incentivised to shore-up the status quo, to adequately interpret the root of the grievances – class inequality – which laid the foundations for Britain's chaotic withdrawal from the world's largest common market.

The discourse around immigration and Brexit acted as a diversion from the fundamental issue. One which many across the political spectrum seemed happy to avoid confronting – immigration and immigration concerns, while seemingly oppositional, are driven by the same economic forces arising from a system of globalisation configured to serve the interests of the middle and upper classes.

While many of the claims made by right-wing populists and their allies in the press in the run up to Brexit were pie-in-the-sky and have since been proven as such with the passage of time, Remainers too were insulated within their own parallel fantasy. The fantasy of cosplaying as continental Europeans. The fantasy that they were somehow kinder to migrants – the EU's response to the refugee crises at times was nothing short of brutal and many Remainers, by virtue of their social class, do not reside in communities where migration occurs on any serious level. And the fantasy that the privileges conferred to them by their class position – freedom to holiday and send their children to be educated or work abroad, and relative protection from disruption to their employment, lifestyle and cultivated identity – are not exclusive. These were delusions the most ardent Brexiteers couldn't hold a candle to and are perhaps even more ludicrous, given how regularly we were all reminded that EU support was higher among the 'educated'.

Perhaps the fury and subsequent wrath of Remainers – admittedly a formidable spectacle once unleashed – lay not simply in the fact of Britain leaving the EU but in the shock and indignity at having been defied – and by those who they privately regarded as lesser. Having become so accustomed to the 'grown-up' politics of New Labour and largely holding their noses (or turning a blind eye entirely) to the media and government obfuscation of the banking crisis, their values and interests remained paramount to successive administrations no matter the political weather. Suddenly, this previously pivotal demographic, so often central to the direction of travel in Britain, was being effectively bypassed, spoken over and even monstered. Had as many middle-class

Remainers mobilised so enthusiastically against the bankers in the aftermath of the financial crisis, or against the Tories when brutal austerity was rolled out, as they did before Brexit, we may well have avoided the constitutional crisis in the first place. Remainers to this day still point to every instance of industrial decline as proof Brexit was a mistake. They implore Brexiteers to realise the error of their ways. But all this does is misread why many voted for it – their lives were already so difficult they felt they had nothing to lose.

For working-class people, Brexit was a false choice between two types of elite hypocrisy. It was a turf war between two rival visions of the world which arise from the same fundamental sickness. The publication of the Pandora Papers in 2021 revealed Britain was not merely implicated as a player in a parallel economy of tax havens and loopholes, accessible only by the wealthy and well connected, but was indeed at its very heart. That no political leader thought to comment on the scandal spoke volumes as to the vapid nature of mainstream politics in Britain – and anyone who thinks that's a terribly populist thing of me to say shouldn't be surprised when the next revolt occurs. The asymmetry in how rules are applied to people of different social classes is proof that the populist refrain of there being 'one rule for them and one rule for the rest of us' is absolutely, 100 per cent true.

The common theme where populist revolts are concerned, wherever they emerge, is the perceived abandonment of the working classes by the political classes. In Britain, this began under Thatcher but was accelerated by the Labour Party, led by Tony Blair. While Blair's pragmatic approach, which we will explore in the next chapter, involved fusing aspects of left and right, social justice and free markets, is often cited as an unavoidable strategic manoeuvre that made New Labour electable, Brexit represents the unforeseen cost when slick politicians attempt to build vast, unsustainable electoral coalitions without a frank parallel analysis of the conflicting social, economic and cultural forces which underscore them.

What hope is there of staving off populist sentiments which naturally arise from inequality when even the left-leaning liberals, who yearn for the good old days of 'grown-up' politics, cannot bring themselves to offer a sufficiently honest appraisal of Britain's hideously configured class system? What incentive does their collective silence provide the more crooked elements of the political class, passing as they do through the revolving doors of public office and the special interests they so willingly serve, to behave any differently? And what hope is there of unifying a divided nation when the affluent and the wealthy in so many ways derive their sense of security, prosperity and superiority from an economic orthodoxy which goes beyond the basic capitalist principles of competition, innovation and supply and demand, and which now so clearly hinges on the coercion, exploitation and political exclusion of working people and the poor?

New Labour is the band many want to get back together, based on the understandably seductive but rather rose-tinted notion that Britain was a kinder, saner nation back then. This reveals not only the short memories of many but the utter lack of proximity the revisionists have to social reality. Many of the trends which ran their natural course, arguably culminating in Brexit, were set in motion when the Blairites rose to power. New Labour is famed for essentially preserving Thatcher's legacy and she herself cited them repeatedly as her greatest political achievement – a facetious but nonetheless keen observation. We are encouraged to adopt the false belief that New Labour was a revolutionary force. That there is a great world of moral difference between the populists, who play to the unjustified resentments of the crowd, and the 'grown-ups' who took the tough decisions whether they were popular or not. But many of New Labour's less publicised policies, when examined objectively, read today like a Christmas list compiled by the most demented, reactionary members of the Conservative Party. While I do count myself among the few voices on the left who concedes some things did get better in 1997, it is also sadly true that far too many things either stayed the same or got worse.

In Britain, this brand of 'grown-up' politics to which many would return in a heartbeat is the rather patronising notion that you can create a fairer society without ever having to speak truth to power. That you can lift the poor without afflicting the wealthy. Indeed, it is the ludicrously self-serving notion that Britain's systemic social inequality can be confronted calmly and politely, without anyone having to embarrass themselves by using the c-word – class. And the populists are the fantasists, apparently.

19

Everything in Moderation
Blairism and the rise of the paternalistic middle class

As a working-class writer, you are often commended by middle-class readers for your 'brutal', 'harrowing', 'unflinching honesty'. While any praise whatever is welcome, I have come to suspect that much of it may be conditional on a number of caveats. As a social commentator with a mainstream platform, you are permitted by the unwritten strictures of middle-class sensibility to offer as spirited a critique of the system's excesses as you can muster but you will soon develop something of an intuition that your criticisms must remain within acceptable bounds, or consequences will follow.

The consequences are not quite as immediate or terrifying as those you may experience for grassing. But there exists, in the middle-class collective psyche, a palpable informal social pressure. One which functions rather like the ideological superego so burdensome on the left or the religious adherence to tradition seen on the right. Nobody has ever sat me down and explained this custom but I have inferred absolutely it exists, not only from the distinct absence of radical voices the further up the rungs I have proceeded but in the relative ease of my ascent whenever I have consciously (or unconsciously) sought to conceal what truly drives me as a writer – anger.

Ascending the structure of Britain's class system hinges not simply on the modifications you might make to your speech style, your appearance and your readiness to perform the rags-to-riches party tricks. Nor is it simply contingent on good fortune or those rare occasions your hard work and talent pay off. Your rise from the working class also rests, perhaps more decisively than any other factor, on your willingness to water down or gentrify any truly radical political convictions you may harbour – or cast them aside altogether.

It would seem, to me at least, that part of the deal, where getting up and getting on is concerned, is that you must buy into the cultural delusion that there exists a route to society's radical transformation which does not require anyone to do or say anything radical. Which brings us to the penultimate chapter, where we will focus squarely on the role of the self-styled political moderate in the unfolding saga of twenty-first-century Britain.

From its outset, the New Labour project was far too keen to keep the middle classes and the rich on side by expanding on Thatcher's reforms to produce anything but temporary progress. And while I differ from many on the left, in that I do not see what else Labour could have done back in the nineties where gaining power was concerned, it is clear the modernisers lacked the courage necessary to truly reform the country. They prided themselves on their cunning strategies and the skilful means by which they were deployed, but is telling wealthy people what they want to hear, even if what they want to hear is bullshit, really that commendable? Having gained power, rather than leverage it to reorder Britain's economic structure for the benefit of all, the Blairites instead squandered their political capital and the faith that millions of Britain's working class had shown them doing the bidding of wealthy and powerful interests for short-term political gain and favourable treatment by the press barons, at the long-term expense of ordinary people and social equality. Indeed, many of the policies they now cite as evidence of their radicalism were enacted by stealth, so craven they were in the face of criticism from the media for acting in the interests of the poorest.

New Labour had as big an opportunity as has ever fallen to any political generation to truly reform British society for the better. Instead, by performatively uncoupling from its own history, New Labour became a hostage to fortune, living in fear of the right-wing press, and ultimately painted itself into a political corner where the betrayal of working-class interests became inevitable.

At the dawn of the twenty-first century, the United Kingdom reached a milestone in its socioeconomic development when the middle class, for the first time in history, outnumbered the working class. It was a landmark moment. One that proved, for many, that Britain, historically riddled by bitter class division and struggle, was finally turning a corner. It was a vindication not simply of the New Labour government, which had swept to power three years prior on a wave of expectation that it would transform the country's socioeconomic fortunes, but also of the economic doctrine that preceded and necessitated its rise – Thatcherism.

Thatcherism had radically transformed the UK's economic and political outlook, wresting power from the state, local authorities and trade unions and placing it in the hands of individuals and the private sector, whose liberation and prosperity would be found in the free market. Thatcher reset the status quo both politically and economically, rendering the post-war consensus, with its commitment to the Welfare State, nationalised industry and closer regulation of business, deeply unfashionable.

There was 'no alternative' we were told. Any divergence from Thatcher's holy trinity of restrained government spending, privatisation of key national industries and decisive marginalisation of trade unions was regarded as reckless, economically illiterate or dangerously radical. This presented the Labour Party with a problem. If it were to stand any chance of gaining power in the nineties, it had no choice but to evolve. No longer could it appeal exclusively to the working poor, the dispossessed or the socially excluded. With the middle class growing, and previous electoral logic changing, the only route to power for the centre-left, it was thought, was to speak to a broader range of social and economic concerns, interests and aspirations than Labour had done previously. Labour's modernisers would accept the fundamental tenets of Thatcher's reforms and rebrand Labour as a party of aspiration, wealth creation and free-market economics – as well as social justice – in an effort to appeal to voters of all social classes. It was hoped that this strategy of moving economically to the right would help convey to the British public (a term New Labour often used as a synonym for the owners and editors of the right-wing newspapers which largely shaped opinion) that it was no longer a party of protest run ragged by trade unions, but a serious governmental contender.

This strategy rejected the age-old notion that politics was ideologically binary or defined largely by a historical struggle between the rich and the poor. New Labour would govern, if elected, by skilfully reconciling the competing interests of the working and middle classes with policies that appealed to a broad range of political sensibilities, while making the necessary overtures to big business and mainstream media which, by then, held the keys to political victory.

'Tough on crime, tough on the causes of crime' was New Labour's early pitch, as it made a beeline for what it regarded as the centre ground of British politics. This slogan was the sort of painstakingly focus-grouped phrase for which Blair and his generation would become infamous. To this day, it remains the primary example of New Labour's clever and manipulative political style. The phrase made one of its first appearances in an op-ed written by Blair himself – then shadow home secretary – in the *New Statesman* in 1993 and signalled Labour's dramatic departure from the traditional leftist rhetoric that had come to define it. New Labour would wrap the iron fist of Thatcherism in a rosy velvet glove, tempting voters in from the right while banking on its traditional strongholds – decimated by a needlessly aggressive process of deindustrialisation – where voting anything but Labour remained a political taboo.

With respect to criminal justice, Blair understood that self-responsibility ranked highly not just in the traditionally Conservative hierarchy of concerns but also among many traditionally working-class Labour voters, whose economic interests aligned more closely with the Left but whose social and cultural values tended to the right as their communities slipped further into crime, addiction and dereliction. In truth, they had grown tired of drug dealers, drunks and thugs dominating their communities.

Blair sold himself – and New Labour – as a common-sense mediator, benevolently intervening to negotiate truces between rival political outlooks. With his undeniable charm, a barrister's persuasiveness and crystal-clear communication skills, Blair would broker compromise by recognising, methodically, the strengths and virtues in what were seen traditionally as competing value systems, weaving them carefully together, producing a new political tapestry in which diverse and lucrative sections of the electorate could see their values reflected back at them. This approach was how Blair and New Labour manoeuvred themselves out of the looming shadow of Thatcherism and went on to redefine British politics in the early twenty-first century. Unlike Thatcher, Blair possessed the ability to communicate with seemingly disparate, previously oppositional sections of the British population simultaneously, while making each feel he was speaking exclusively to them. It was a parlour trick. A dance of sorts. A carefully constructed political performance – and he was really good at it.

At the turn of the millennium, British demographics were changing. While Thatcher's policies had led to poverty and social dysfunction in those communities which had depended almost exclusively on the nationalised industries of coal and steel, others benefited immeasurably. Thatcher is still held in such high regard by so many for this reason. Many believe the negative impacts felt by post-industrial Britain were a necessary though painful sacrifice in the broader quest to unlock the country's economic potential. Others privately understood that inequality would be a consequence of neoliberal reforms but that the net benefits of wealth creation, innovation and competition made it a price worth paying – a conclusion you can only reach when you are not directly impacted.

By the time Blair came along, the expansion of the middle classes created by Thatcherism had fundamentally changed the nature of politics. The electoral logic was simple: appeal to the affluent and you stand more chance of winning an election. Win an election and you have more chance of helping the poor. This route to power was regarded as 'pragmatic' because it recognised the need to appeal across traditional party lines. Not only did it appear to make practical strategic sense but it also gave rise to a perception that previously competing concerns could be reconciled in a manner that would make society more harmonious and cohesive. That Britain could move beyond the class divisions and struggles which had, in many ways, shaped and defined it.

As more individuals and families ascended to the middle class throughout the eighties and nineties, their political and economic needs and personal and professional aspirations changed – as did the need for politicians to consider much else but what the middle class desired. As incomes stagnated, credit became available for all manner of purchases, allowing households which on paper had not ascended to the middle class to create the illusion of security and prosperity. The abundance of credit made many feel they were moving up in the world and they attributed this either to Thatcher's embrace of their aspiration or Blair's recognition of their merit – not the financial sector that would later enslave many of them with debt. The fact almost all politicians, from leaders to backbenchers, were by then also middle class made this new reality all the more intuitive and, as more people ascended, middle-class interests came to preoccupy and dominate British political culture. Typified, perhaps, by Blair's famous: 'Doesn't everyone want to be middle class?' observation. A comment which betrayed his lack of insight into the consequences of a society which is dominated by one social class.

While I do not hold with the view that nothing changed under New Labour, what change did occur was to be short lived. In essence, the policies it adopted to get elected and the constraints these then placed on its ability to radically reform Britain's institutions, compounded by a collapse in public trust over Iraq, would become its undoing. But even in its heyday, whatever gains working-class people and the poor experienced economically were undermined culturally by an aspect of the Blairite period which is rarely noted.

Blairism found expression culturally, in the longer term, in the infantilisation and caricaturing of the working class and the poor as vulgar, feckless and workshy. Some may argue this culture was a media-generated remnant of Thatcherism but a closer examination of New Labour's record – what they did and what they tried to do – tells a different, less flattering story. Benefit cuts to single mothers and refugees, 'name and shame' orders designed to deter anti-social behaviour and bizarre plans to kick council tenants out of their homes for being unemployed are not often associated with New Labour. Nor are claims that certain ethnic groups 'lacked discipline', clampdowns targeting 'foreigners who come to this country illegitimately and steal our benefits' or rather fantastical claims that translating basic public service information for those arriving in the country from abroad meant they would 'not have the incentive to learn English' (these quotes taken, in order, from Tony Blair discussing knife crime in London, home secretary John Reid talking about benefits on BBC

Breakfast and communities secretary Ruth Kelly on English language and integration, all from 2007). These things are not associated with New Labour in the popular memory but that doesn't change the fact they happened.

A conception of poorer sections of the population as untrustworthy scammers looking for a free lunch took hold. Negative portrayals of working-class people and the poor in news and entertainment media led to the formation of new and dangerous false beliefs both within the working class about itself and across wider society. The highly publicised ASBOs and active Welfare State agendas, *The Jeremy Kyle Show*, *Little Britain*, reality TV generally and the rise of mean-spirited television presenters like Anne Robinson and Gordon Ramsay, whose shtick often lay in their goading of working-class participants, normalised the notion that it was open season on the lower orders.

While Blairism is often associated with a style of politics which involved slick media presentation and spin, and reforms which led to temporary reductions in child poverty and increased social mobility, it is also characterised by its cultural offshoot where working-class life and working-class people were often regarded as coarse, suspect, dangerous or childlike. This in many ways set the table for the ascent of the middle-class paternalism which, in my view, defines Blair's tenure domestically. The managerial middle class would derive new purpose by directing, chaperoning and compelling working-class, poor and vulnerable people to embark on journeys of self-improvement, based on the popularisation of powerful myths which attempted to explain the widening gulf between the classes. Social mobility, the poverty of aspiration and meritocracy all but supplanted traditional class analysis and in their reification class consciousness evaporated. What's more, many middle-class progressives who had previously opposed Thatcher's economic policies would eventually turn something of a blind eye to New Labour's commitment to affirming them – Blair was elected in part on the basis of a promise to stick to the Tories' public spending plans. Had the left-leaning middle classes not been offered such a central role in managing social inequality they may have proven harder to co-opt. But ultimately, Blairism was reluctantly co-signed by many in public services and the Third Sector because of the opportunities provided by the burgeoning poverty industry, which would be unleashed to whip the plebs into shape.

Without the language of class to orientate itself, the working class was once again segregated according to middle-class interests, assumptions, and organisational and party-political incentives into the age-old categories of deserving and undeserving. Worthy and unworthy. Useful and useless. With Blairism, rather than a classless society, came a new and brutal kind of class war. Not one waged aggressively but through inauthentic, middle-class politeness. From the classroom to the court room, the social mobility of those of the lower orders would depend on the approval or assistance of a middle-class broker. A middle-class advocate. A middle-class gatekeeper. A paternal figure imbued with the authority to corral and compel you to pull your socks up and speak properly. The aim was no longer to remember where you came from should you have been lucky enough to make it out but to cast aside (or learn to conceal entirely) all traces of your working classness. To redeem yourself, in the eyes of the middle class, you would have to learn the steps of their dance.

As the middle class expanded further, it began flooding working- and lower-class communities through arts initiatives, back-to-work programmes, careers advice and mental-health services, and alcoholism and addiction counselling, bringing with it a cultural sensibility which, while certainly not homogenous, was distinct enough from working-class culture that many found it patronising and jarring. The emphasis on politeness. The aversion to profanity. The constant commands to 'calm down'.

Community dynamics that once occurred naturally in lower-class communities when there were opportunities to exploit and social connections to forge and nurture would be re-imagined, but as a free-market simulation of community, mediated by 'facilitators', 'mentors' and 'practitioners' who would 'model' new ways of being. Lower-class people would once again come together, but not without a middle-class professional to carefully choreograph them.

Even as our communities improved aesthetically, often through gentrification sold as urban renewal, new tensions arose as class conflicts began to evolve in unforeseen ways, which were difficult to identify and articulate in the absence of the old language of class.

When a new school was built, that's what made the local news. What you were less likely to hear about were the numerous public meetings at which local opposition to the new school was voiced. When a shopping centre was superimposed onto public space, and often given an utterly contrived name to disassociate it from the 'notorious', 'deprived', 'poor' community in which it was set, this mall, which offered locals low-paid, insecure work while catering to the high-end tastes of suburbanites a few junctions down on the motorway, became emblematic of our communities' new success stories. We were edited out of our own histories. The deep sense of exclusion and grievance voiced by people in those communities was often dismissed, muted or pacified. Working-class people just didn't see the bigger picture. They weren't smart enough to know what was good for them. The choice for the lower classes was abundantly clear: you can have nothing under the Tories or you can have middle-class people managing your communities with very little authentic input from you.

This so-called 'poverty industry' is a conveyer belt of initiatives, projects and engagements often riddled with the assumptions of those who have little proximity to the lives of those they seek to control. This industry now represents a significant plank of the UK economy, providing an endless stream of professional opportunities for the aspirational middle classes – and stands as just one mechanism by which they now dominate Britain. Indeed, poverty, adversity and political exclusion are its main fuel sources. Without a disenfranchised lower class to advocate for, and speak on behalf of, countless organisations and professionals would quite simply find themselves on the scrapheap. This inversion of dependency, where value is extracted from one social class and appropriated by another, reflects the heartbreaking marketisation of community life and democracy – New Labour's true legacy.

The enduring appeal of centrist politics is not that difficult to understand where the more liberal middle classes are concerned. It was assumed, with good reason, that the more people ascended to the middle class, the better the economic fortunes of the country. But as the middle class has grown and come to outnumber the working class by an increasing margin, the scales have begun to tip a little too far in favour of those who possess assets, such as property, who do not depend on public services or social insurances to the extent working-class people now must. And so, politics is now dominated by the need to appeal to the values, aspirations and interests of the top 20 per cent of the income distribution, which are largely taken care of, irrespective of whether they vote Labour or Conservative – Brexit being a very rare exception.

New Labour's (and by extension centrism's) original sin lies in its fealty to power. The forces which shape public opinion and dictate how our economy is structured. Centrists make this concession under the guise of pragmatism, proclaiming to accept the world as they find it, and then commencing to marginalise all who do not see this as a viable long-term strategy. They refer to public opinion as if it arises naturally in a vacuum and is not, in part, a product of a great many vested interests – odd for an outfit so rightly obsessed with media. Centrists become so invested in this notion of themselves as skilful strategists and mediators that they cannot truly confront the negative outcomes of their Faustian bargain and have yet to create a diagnostic tool, or identify some basic threshold, for assessing their own failures. They regard calamitous events like Brexit as symptomatic of the absence of centrist politics and not partly as a consequence of it. When their denial runs that deep and their lack of humility is so apparent, can we really describe such a stubborn resistance to reality as 'grown-up'?

I understand the attraction to Blair. He was and remains a gifted communicator and possesses a wonderful analytical mind. New Labour, in the beginning, was a substantive political project, and this cannot be denied. The problem today is that moderates refuse to see the lessons so clearly staring them in the face. Instead, they desire a return to the New Labour heyday, despite misremembering exactly what occurred.

They confuse the undeniable effectiveness of New Labour's slick public relations operation, and Blair's unparalleled skill as a political performer, with far-reaching social progress when the real tragedy of this endeavour was that all of those wonderful tools and skillsets, and so much energy, time and thought, were deployed in service of an agenda which, at its root, did not diverge greatly from the Thatcherism voters had just rejected. The immense political capital at its disposal in 1997 was sufficient to undertake a far more ambitious programme of public sector reform than was ultimately mounted. New Labour possessed the talent, discipline and confidence required to revolutionise Britain as Thatcher did, perhaps with even more public goodwill than she had ever enjoyed. Instead, they played it safe and, in doing so, planted the seeds of their own political destruction.

Blairites understandably point to the minimum wage, Sure Start, smaller class sizes, school investment, the shortest hospital waiting lists for 40 years, 11 years of uninterrupted economic growth, devolution, peace in Northern Ireland and a largely constructive relationship with the European Union as proof they were distinct from the Tories. But in every instance, this progress was built on soft sand. Inequality has risen. School investment has fallen. The NHS is on its knees. The economy is dysfunctional. The Union is coming apart. Peace in Northern Ireland hangs-in-the-balance. And a largely constructive relationship with the European Union is in the toilet. Meanwhile violence is a daily occurrence in Baghdad, and Afghanistan has been retaken by an invigorated Taliban. The only remnant of Blair's tenure that seems permanent is the debate about the man's character. You couldn't say the same about Churchill, Atlee, or Thatcher – their legacies are woven into the fabric of British life, for better or worse.

Blair remains an immensely influential figure among moderates on both sides of the aisle, perhaps even a de-facto leader for some to this very day, using his Institute for Global Change to frame everything from the pandemic response to the technological revolution, Corbynism and even his own legacy for the benefit of those still in the field, who look to him for guidance and instruction. He subtly steers the right of the Labour Party, with strategic media appearances and regularly published policy analyses which generate headlines, commentary and debate. And every time he does, he elicits a warm, comforting emotional response in those who sorely miss his presence. In many ways, the repeated efforts of moderates to downplay his failures as a leader or, worse still, to attempt to politically rehabilitate him, speak to the painful lack of proximity the self-styled 'adults in the room' often have to the truth of their own motives. Blair is a political father figure whose interventions they find reassuring because he always sounds like he knows what to do in an age where top-flight politicians appear so out of their depth. And this speaks to the delusions of those who now claim to be politically homeless in the absence of such a transfixing figure – they long for a politician whose rhetoric reassuringly obscures the class conflicts and inequalities which entrench their economic advantages and cultural sensibilities.

Centrists in the New Labour mould, whether centre-left or centre-right, want everything both ways and no longer recognise that sometimes we need to take on the powerful – not jump into bed with them for an easier life. Moderates want change but things to stay the same. They want debate but within certain limiting parameters. They want to help the poor but see the world through the eyes of the affluent and the rich. The centrist delusion that they are not driven by ideology has provided their managerial fundamentalism with a convenient justification for possessing no discernible moral core.

Whether centrist politicians or 'politically homeless' voters, lamenting the good old days, invariably, these are people who have prospered. People who possess the capacity to create opportunities merely by picking up a phone. People who might pay more tax or might pay less tax but whose general interests, aspirations and 'values' will always be protected. People for whom politics will never be a matter of life and death but merely a vehicle by which they can express a personal preference. A window in which they may exhibit their paper-thin values, while lecturing others about being civil and polite.

It's easy to be calm and civil when political decisions don't send your life into a dangerous tailspin. It's easy to be polite when you haven't lost countless loved ones to deaths of despair. Of course, this condition that one must conduct oneself graciously, adopting middle-class mannerisms, is also disingenuous; civil debate is not simply about being graceful and keeping your cool but is also key to the maintenance of politically diverse social networks on which career opportunities and social prestige depend – that's why moderates always perform a pandering flinch or look for a middle ground in the face of a problem which may involve genuine conflict, sacrifice or steadfast commitment to principle.

Their coveted 'consensus' is simply the location where everyone with a stake in the ailing status quo meets to occasionally decide what they are not prepared to give up to help the millions still trapped in the quicksand. Moderates, in the face of the disturbing levels of inequality we see, are nothing short of militant in their commitment to an economic system where wealth does not trickle down but accumulates at the very top. They cling desperately to the dream that this will all blow over. That things can only get better. And to keep this fantasy alive, they deploy every form of capital at their disposal, whether financial, social or cultural, to reinforce the walls of their simulation, undergoing all manner of hideous political mutations, revising the narrative as they go, so that they never have to concede how repeatedly wrong they have been on the key issues facing the UK since the turn of the millennium. In the process, their pipedreams become the nightmares of the less fortunate, as moderates work round the clock to create and sustain a carousel of political and media sideshows as a poor substitute for Britain's overdue and unavoidable reckoning with class inequality.

The centre ground is where good people, who were once idealists, openly brag about casting aside their principles while sneering at and marginalising those who retain them. The centre ground is where good people are bought off, flattered and pampered into moral submission, before professionalising their class interests and retrofitting the compromises they made to their basic integrity on the way up with self-serving narratives about pragmatism, merit and maturity. Adults in the room? Grow up! For surely the hard truth is this: when it comes to poverty, Britain's liberal middle classes are not as interested as they like to think they are in 'brutal', 'harrowing', 'unflinching honesty'.

Coda
Closing remarks

After a long day of commuting, I make the mistake of coming up for air at Oxford Circus, thinking the busy retail district may have calmed slightly, given it's nearly 8pm. As I ascend the escalator and emerge into the station, engaging in the uneducated guesswork all non-Londoners must when navigating the tube, I am finally overwhelmed.

Attempting to sync with the heaving mass making its way down Oxford Street, I long for home. Barely able to hear myself think over the noise of cars, buses and indiscernible accents, my eyes fall to the pavement. To my left, sat in a pile of newspapers and cardboard, sits a young man with his hands over his ears, audibly sobbing. He is begging, but not for money. I speculate that he may have a sensory perception condition like autism spectrum disorder, and his distress could be a result of the sheer volume and movement going on around him. But before I've even had a chance to observe him, I am shuttled along the great conveyer belt, his cries becoming fainter as I am pulled away and back into the consumer melee.

Not one minute passes before I see another, this time an older man with an artificial leg. Then a woman with a dog, painting on some carboard. The streets are literally lined with beggars and rough sleepers. After a time, I become irritated by them all for protruding on my retail fantasy and decide to cross the road, into the Nike store on the corner, where the sound of hip hop music anchors me temporarily.

After rummaging around, across five floors, I settle on a grey hoodie. 'I deserve it,' I tell myself. 'I've worked hard.' I leave, feeling accomplished and looking forward to trying on my new garment. I am then confronted by a sign-waving preacher. His placard reads:

'Jesus Saves Us From Hell'. I think to myself, that's a great pitch for a taxi company that gets you out of central London in minutes for a flat fee. He then turns the sign 180 degrees, revealing another message: 'Repent Or Perish'. In a sea of thousands, I can't help but feel personally attacked as I place the trusty headphones over my ears and plummet once more into London's turning bowels.

As I make my way back to the flat where I am being put up, in the district of Pimlico, a ten-minute walk from the north bank of the Thames, where a man could walk around for hours without ever feeling threatened, I realise I am, to borrow a phrase, further from home than I have ever been. This was not how my life was supposed to be. I was supposed to be a firebrand and a radical. My life should have been stubbed out early by a gratuitous act of violence or an overdose. I am not supposed to have money or nice things. Yet, here I am, heading to the London flat owned by my publisher, ahead of tomorrow's meeting to discuss my next move up the chessboard.

Suddenly, my world has opened up, the possibilities seem endless, but I cannot shake the certain knowledge that things are getting worse for those I fear I may be leaving behind me. I feel the immense gravity of that promise of prosperity bearing down upon me, pulling me further from the world I know. The world I understand. The social ties to the communities I love are becoming frayed or have been severed entirely. At least, that's how it feels. I struggle to relate to people I once counted as my dearest friends and conceal much of my business from family. I still have a sense of what it's like to struggle but it washes over me like a wave of nostalgia rather than a painful memory. My claim to being working class is becoming more tenuous by the day, as is my grasp of who I am and what I believe. I fear I have been corrupted.

I approach the building, stubbing a cigarette out on a wall and throwing it in a bin, and the concierge opens the door before I even reach the handle. I am truly a fish out of water. I catch him scanning me up and down, probably thinking to himself that I do not look or sound like I belong here. I step in the lift and see my reflection in the mirror, before letting out a bewildered sigh of resignation. I don't.

This might surprise you, given how often I go on about it, but I don't enjoy talking about class – especially these days. It wasn't always this way, of course. It used to be fun to talk and write about it when I was struggling. Struggling to hold down work, struggling with mental health problems and struggling to get and remain sober. I ran almost exclusively on anger. Class was the indiscriminate fist with which I would righteously 'punch up', much to the satisfaction of my comrades.

Well, times have changed. For me, at least. It's far trickier to touch the topic of class when you've made a few quid. That is, if you are blessed with self-awareness. When you're no longer scraping to get by, questions are rightly asked as to how authentically you can represent working-class experiences. Having spent my life railing against middle-class people and the wealthy, certain that their affluence was in some way linked to my hardship, I now find myself at an uncomfortable crossroads. In the past few years, I've been given a taste of how the other half live, as the material circumstances of my life have changed dramatically.

I still feel like the same person. I still feel working class and I like to think I remain close to working-class concerns. My cleaner is working class. My personal trainers are working class. In fact, my old photographer was so working class he couldn't even get backstage at my Edinburgh Fringe show without first having to aggressively refute an assumption that he must be a racist.

I made my name writing about poverty and, as a result of the spoils from that labour, now find myself in the top 8 per cent of earners in the UK. Even a bad year, when a pandemic hits and the arse falls out of the economy, leaves me only slightly flustered. What I have observed is that there is some truth to the old adage that 'money goes to money'. Wealth is like a magnetic force, even the moderate kind I have experienced, drawing to it countless opportunities, both personally and professionally, which not only generate more money but also afford me a higher quality of life.

I'm physically healthier than I have ever been. My debt burden is significantly lower, as a portion of my income. I'm also in a position to support my immediate family in times of need and channel resources to relieve the various stresses that I once regarded as an unfortunate fact of life. After many years of feeling (whether right or wrong) socially and culturally excluded, people listen when I speak. Indeed, I am overwhelmed with invitations every day to do just that. The fight to be heard is over. It would seem, to anyone looking on from the outside, that I have, despite the odds stacked against me, 'made it'.

Some people seem bred for success. When people of a higher social caste become successful it is unremarkable and rarely something they must justify. When you grow up in poverty and become successful, you are never allowed to forget it. Few would ask privately schooled Glaswegian historian Niall Ferguson where he learned 'all the big words' he uses. And that's not even the worst of it. The real boot in the balls comes when, having been deemed successful by society, you are repeatedly held up not simply as an example of a 'diamond in the rough' but as a poster-child for the system itself. The system you succeeded in spite of. Having made your name railing aggressively against the status quo, you are suddenly celebrated as living proof of its enduring utility.

But one mustn't complain. One must be grateful to have ascended the slippery snakes-and-ladders board of British society, right? Mustn't rock the boat too much, for what is given can also be taken away and I (hubris alert) worked hard for my seat at the table. The fear of losing that seat now plays an intrusive role in my thinking, so utterly bowled over am I by my luck that I dare not protest too much in case the seat is abruptly revoked. Rather than success, prestige or wealth giving me more freedom to tell society some hard truths, it has become a rein which, when politely tugged, brings me reluctantly into line. Kicking back against this, however, is something I have decided I must do.

Being granted a seat at the table, rather than unleashing my inner revolutionary, has had an immediate, moderating effect. Some of you know what I mean, don't you? Those of you who can still recall the moment you were offered a seat. You remember the sudden tension between concerns about your community and concerns about yourself – your earnings, your social standing, your legacy and your next move up the board. When you began to see yourself as your colleagues, contemporaries or an audience might see you, and your incorruptible principles developed a sudden and convenient elasticity which, rather than proof of a forgivable lack of moral fibre, became evidence of your fair-minded maturity. The previously strident views you once held and regularly espoused became easier to laugh off as naive and idealistic. The jobs you called other people hacks for doing gradually became desirable. Those expense accounts you once criticised as excessive were suddenly justifiable and fair. You changed. You adjusted not only to a certain quality of life but also became intuitive to what it simply would not do to say, should you wish your run of luck to continue.

It's amazing how quickly you attune to the prevailing sensibility in this lucrative domain of legitimacy. How grateful you are for the exacting dimensions of the professional sandbox you have been permitted to play in. I'm now the guy who talks about poverty. Sometimes I fear my seat at the table is contingent not on a broader analysis of inequality or politics such as I have attempted in this book, but on my willingness to regurgitate my own 'lived experience' of class, addiction and trauma. I still harbour radical beliefs. I'm still angry. But every seat at the table comes with its own unique restraint. Mine is no different.

I can still vividly recall the struggles I faced prior to moving up in the world, but that visceral sense of poverty's all-consuming immediacy is impossible to retain. I am already beginning to forget what it felt like to struggle on a daily basis. The impact of political decisions on me is now negligible. That's why many in my tax bracket can afford to lecture others about being 'civil' and 'polite'. Affluence pampers you to the extent that you can no longer relate to those whose lives are routinely thrown into chaos because of ill-considered (or downright malign) political decisions. Three or so years of living comfortably has partially numbed me to the reality millions face in the UK, so what hope is there for those with real power and influence, who rarely contend with that reality at all, or those who wish to contend with it but possess only a theoretical understanding, to bring about meaningful change?

In 2015, when I started my first book, the situation in the UK did not seem quite as bleak from where I was sitting. The full extent of the UK government's austerity measures, immigration reforms and inaction around the drug-death and homelessness crises had not fully materialised. There was no 'Brexit' dominating public discussion, social media and the news agenda, and the notion of Trump being elected was laughable. I began writing my first book when the world we live in now was something most people would never have dared to imagine could manifest.

In that book, among other things, I made a rather controversial call to my comrades that we would require more than just anger to reorder society. That while systemic change was obviously necessary, the fundamentals of our society remained sound. But by the time that book was published, this nightmare we now call reality was beginning to take shape. Suddenly, there were so many issues emerging that warranted analysis. From poverty to immigration, austerity to criminal justice, and, not least, looming on the horizon, the biggest peacetime political crisis Britain had ever seen. I began to feel like I had sold my politics short. Worse, that I had sold my community out and not even realised it.

Then, people I knew started killing themselves. Becoming homeless. Using food banks. And I started getting angry again. I started getting angrier than I had ever been. But, to my surprise, I was also beginning to experience security, affluence and prosperity. And then it hit me, right in the guts, why there is so little appetite for radical change in the UK: the system advances the aspirations and interests of just enough people that our most influential social class hasn't yet become willing to accept the fact that our way of life will have to change radically – whether they like it or not.

The Covid-19 crisis, as many have stated repeatedly since its emergence, merely accelerated a process which was already well underway – the political, social and economic decline of Britain. While the ominous cloud of coronavirus continues to lord over us all, casting its unending shadow over our lives and livelihoods, the true character of British society has been laid so utterly bare that, in the years to come, the painful revelation may prove to have been the slenderest of silver linings.

So, what have we learned? Well, we learned that class remains the primary dividing line in society. This was evidenced early in the crisis when half the country moved onto Zoom, while the other half delivered them alcohol, sex toys and bread-makers. This was evidenced in the higher rate of infection in communities that depend on public transport, struggle to access healthcare services, where housing is of a lower quality and too many people are crowded in too little space. This continues to be evidenced in the level of exposure of essential workers not simply to infection but also to unemployment, mental health crises, the grief of loved ones dying and to perishing themselves.

We learned the state interventions are actually OK – as long as they are targeted disproportionately at the livelihoods of economically viable, politically lucrative, over-mortgaged sections of the population, whose 'stay-cations', credit cards, gas-guzzling cars and gym memberships must be covered at all costs. We learned that some people are trustworthy enough to receive non-means-tested state handouts by the tens-of-thousands while others must wait an arbitrary number of weeks for benefits that are not enough to live on. We learned that after a decade of austerity (which left tens of thousands destitute, homeless, hungry or dead), the 'magic money tree' we were told didn't exist was not only real but had blossomed so prolifically that the state was able to fork out millions providing discount Nandos over the summer. We learned that 'community' and 'looking after your neighbour' are more important than individualism and self-interest and that when politicians use their public platforms to emphasise social solidarity and not division, and that it's OK to regard ourselves as more than rats in a race, the national mood can shift quite dramatically from mean-spiritedness and distrust to compassion, empathy and harmony. We have learned that any crisis which might befall the middle and upper classes will be rapidly escalated to the status of a public health emergency within days but that drugs deaths, rough sleeping, child poverty and femicide, epidemics spanning decades, still do not qualify for this special designation.

We learned that governments 'follow the science' when it is politically convenient but that leaders actively ignore and dismiss the science when it comes to issues that affect the vulnerable, like drug addiction and poverty-induced child neglect and abuse, or human social connection.

You could give every rough sleeper a home with the money the government wasted on things that didn't work. You could reduce the attainment gap for a fraction of the cost of the furlough scheme. You could cut child poverty and drug deaths by thousands by closing off the tax loopholes that allow the wealthiest and best connected to squirrel away billions. Before this pandemic, people would have called you a loony for suggesting – after the financial crash for example – that the government should step in and just pay people's wages until a crisis blew over, but when the interests of the affluent and the wealthy are threatened, what was once deemed dangerously radical suddenly becomes a simple matter of common sense.

As is customary with any public airing of grievances, one must also provide a shopping list of solutions. While I am still slightly daunted by the prospect of this (perhaps too adjusted to anything remotely radical being dismissed out of hand) I will nonetheless briefly outline three actions we could take which might ameliorate the worst proximity gulfs in education, the labour market and within our democratic institutions which, when taken together, will bring social classes closer together, create conditions for widening equality and bring those in power closer to the action – and accountability. But, more importantly, these actions will empower working-class people to push back against the worst excesses of the current system and, in turn, to some degree, save the privileged from themselves, by rebalancing Britain.

Transforming society begins with education which, as we established from the outset, remains the primary guarantor of class inequality in the UK. It's in the various education systems across the home nations that children are broadly segregated according to social class and where pathways to further and higher education, as well as the labour market, are set. The consequences of the disparity between pupils on the basis of the wealth of their parents is clear – a child who happens to be born in the right postcode will not only tend to have an easier time of it, with less exposure to stress and adversity, but will, as a direct result of this, go on to achieve higher grades. They will also earn more and live longer than a child of equal ability who is born in a poorer postcode. As a result of this early advantage, a child born to a more affluent family can expect little-to-no police or criminal justice involvement; generally better health as well as better access to superior health services when they do require them; less incidences of health-risking behaviour; less exposure to community or household violence; less likelihood of residential instability or homelessness and less chance of ending up in prison or on long-term benefits. Equality is not about rolling back the advantages enjoyed by the privileged but about recognising how pivotal those advantages are and working to extend them to all children as best we can. Nothing short of an educational revolution will achieve this. One based on the principle that every child has a right to the same quality of education, irrespective of the class position of their parents.

To truly test the theory that Britain is a meritocracy, all fee-paying in education must be abolished and replaced by a fully comprehensive system, defined by equal access, where school allocation is lottery-based. The independent sector, by its very nature, discriminates against children on the basis of their social class by denying access to those who cannot afford it. This concept is simply an affront to the notion that Britain recognises individual hard work or talent.

In Finland, education is free at all levels from pre-school to higher education and while private schools can and do exist, they are forbidden from charging fees and from implementing exclusive selection processes – this is the direction the UK must move in. In pre-school and basic education (which begins at age seven), all textbooks, lunch and transport for children who live outside the immediate vicinity of the school are provided by the state. At upper secondary level, all pupils are entitled to a free meal and even in higher education this is state-subsidised. Indeed, adult education is the only form of education that may require a financial contribution. The basic education in Finland is notable for its commitment to maximising the potential of every learner, regardless of their background, and recognises the importance of relational proximity in teaching – in basic education, most pupils will rise the ranks with the same teacher for a number of years before graduating to the next level.

Young children are graded by continuous assessment, not national tests, and a pupil is not expected to sit a national exam until the age of 16. Pupils with additional needs are entitled to special support within mainstream schools, as opposed to being triaged in remote educational institutions, creating proximity between children of different abilities and backgrounds, who come to view those who diverge behaviourally, academically or culturally as a mark of diversity within their community – not as separate or 'other'. Educational autonomy is high at every level of the Finnish education system, in recognition that while a society's educational institutions must work towards common values, standards and labour market needs, a one-size-fits-all approach can create unnecessary problems as well as proximity gaps between the needs of teachers and institutions (often dictated centrally), who must demonstrate competence and standards, and the needs of pupils, who often require an individuated education, suited to their unique developmental trajectories and social and relational circumstances, to keep them on track. Schools in Finland, for example, may be granted a greater say over what daily education looks like, how lessons are structured, what teaching practices are adopted and even the power to decide class sizes, as well as the power to localise the national curriculum to fit their specific social, cultural, economic or academic context. Most notably, school inspections were abolished in Finland in the 1990s – a recognition that steering schools towards higher standards is far preferable to the remote enforcement we see across the UK.

Ultimately, by placing equality at the heart of education, we pay more than lip service to the notion of meritocracy. By abolishing fee-paying education, we enrich the lives of children from all social backgrounds by allowing and encouraging them to mix. This would lead to greater understanding and kinship between children from different social backgrounds; a cross-pollination of skills, histories and experiences, as well as teaching approaches, which would guard against the formation of middle- and working-class educational ghettos where many of the culture gaps explored previously often begin. This is not about attacking wealthier families or limiting parental choice, but about recognising that a parent's right to use their wealth to send their children to the 'best' school represents an infringement on the rights of less well-off children to as fair a crack at the whip as anyone else. The blind upward distribution of opportunity in the UK over the last century or so has contributed massively to its social, economic and cultural decline. The notion that competence and hard work explain why Boris Johnson is prime minister washes with fewer and fewer every day – his ascent to Britain's political apex is a direct result of the privilege conferred on him by attending the country's most elite school.

Private schools have much to teach us about the possibilities of education when it is adequately funded, where school days are configured around maximising the potential of every student and not around the limitations of dwindling budgets. But perhaps the biggest lesson we might learn from the independent sector before it is wound down is the pivotal role of cultivating and nurturing strong social networks which can sustain a person throughout their entire professional life – the product ultimately being sold at every fee-paying school. I have seen for myself the benefits to pupils who attend schools which are well funded. And have spoken with many teachers in the sector who understand intimately the problems inherent to a two-tier educational system.

If abolition seems a bit too extreme for now (funny how any talk of constraining the privilege of those who already have plenty is so frightening for some while withdrawing school meals from the poorest kids during a pandemic seems a reasonable thing to do), perhaps we might cleanse our palettes by revoking the charitable status enjoyed by independent schools and pegging the funding of the state sector to at least 80 per cent of what private schools generate per head. Currently, privately schooled pupils receive almost twice the amount per head than those in state schools. If we think of money as a vital nutrient (or a performance-enhancing drug), then it is clear the current situation in Britain is tantamount to educational apartheid. Critics of the abolition of fee-paying schools will claim there is no point in pursuing such a course of action as wealthier parents will simply identify new means of giving their children the advantage – extra tuition being the main one. The response to that is simple: if they believe abolishing private schools would make no difference then why have them in the first place?

Finland, as well as being ranked first again in 2021 in the annual World Happiness Report (for the fourth consecutive year), is also regularly ranked within the top ten most equal countries in the world. In the UK, we have fallen prey to the rather utopian notion that a market-inspired approach will increase choice when in truth any choice that does exist is open only to wealthier families. When parents are incentivised to follow only their own individual interests, there is a collective consequence for all of us with respect to education – poorer kids being denied opportunities while increasingly mediocre middle- and upper-class people ascend the ranks of politics, pulling up the ladders as they go. While we may never be able to correct completely for wealth disparity as it is expressed in the attainment gap, the social and cultural benefits of children from different social backgrounds learning side by side will be quite simply immense. Not least as, having actually been exposed to working-class people in a setting where they are equal, middle-class children who do still ascend professions will do so with a firmer idea of the society they preside over, and a greater sense of proximity and solidarity to those with whom they shared a profound educational experience earlier in life. In any event, if individuals really do rise according to their ability and work ethic – and not the contents of their parents' bank accounts – then I see no reason why anyone with genuine faith in their child's merit would have a problem with consigning independent schools to the dustbin of history.

Given the education system's relationship with employment, it would be folly to propose radical reform of schooling without a parallel examination of the labour market it serves. Industrial relations – the patterns of conflict and cooperation between workers and employers – are a very dry topic for many people, not least for the millions of workers who are not represented by a trade union in the UK, who would benefit from being better informed, given how central collective bargaining has been and remains to fairer pay, improved working conditions and greater employment security. Despite a small rise in trade union membership over the last few years, reports from the Office for National Statistics show that overall membership has declined considerably from 32.4 per cent in 1995 to 23 per cent in 2020. This decline has been steady since its peak in 1979, when a new Conservative government came to power and quickly began dismantling most labour rights; a total of ten Acts were passed throughout the decade, which constrained the autonomy of trade unions and the legality of industrial action. While much of this legislation was popular at the time – the previous decade was marked by economic disruption which resulted in greater incidences of strike action with which many citizens became fatigued – the results for workers are clear: wages have not risen in line with living costs and the amount of money the average working-class family needs to live a decent life far exceeds what they can earn even in full-time employment. In-work poverty has become the norm. As with the myth of meritocracy, the fib that a good work ethic is enough to lift yourself out of poverty is exposed.

Tackling social inequality in the labour market, by strengthening worker representation, is essential to rectifying this imbalance. A government which is also willing to repeal some of the anti-union laws which constrain working-class people's living standards wouldn't go a miss either. But whatever shape any potential change might take, it must be driven by workers themselves. Workplace democracy comes from workers organising themselves and thinking strategically in their demands for power. In the absence of this, there's no real accountability. Power can be taken away just as easily as it is granted when it remains in the gift of bosses. Processes run the risk of being tokenistic without critical engagement from workers.

A federal minimum wage (which is not enough to live on) or even a job guarantee won't necessarily cut it; workers require greater proximity to decision-making in the workplace, and greater degrees of accountability to them for those at the top must be baked into the equation. It is well understood that economic equality and workers' rights are linked. As trade union membership has declined in the last 40 years, sluggish pay-growth and the subsequent surge in incomes at the top have created dangerous levels of inequality. Lower levels of unionisation have allowed employers to offer lower pay and poorer working conditions as well as avoid penalties for malpractices, which create immiserating conditions for millions of workers, who are viewed as easily replaceable. Strengthening workers' rights and expanding workplace democracy are issues which are too often framed in politically self-serving terms that suit big business and too little is made of the fact that income equality improves in the presence of worker representation.

As a culture, we have grown dangerously remote from the reality that trade unions are vital mechanisms by which working-class people secure basic dignities. Decades of media hostility and historical revisionism have seen workers' rights marginalised and monstered in equal measure; they are consistently presented as a barrier to social progress rather than a driver of it. And yet, worker representation not only protects workers but also underwrites much of what we regard, and frequently cite, as evidence that Britain is a civilised, developed society: the two-day weekend, maternity and paternity leave, equal pay, more flexible working hours, workplace health and safety, sick pay and pension protection, holiday entitlement, defending public services and workplace discrimination safeguards.

Trade unions predate the Labour Party and exist not simply to jostle for position within it, or to cause disruption through strike action, but guarantee a baseline of equality across the board. It is well understood that in countries where rates of collective bargaining are higher, lower levels of inequality are experienced and that within firms where trade unions are present, pay inequality falls. In recent years, with the rise of precarious work, surplus labour and zero-hours contracts, workers have become vulnerable on many fronts. Indeed, the gig economy has changed the relationship between the individual worker and industry; the peripatetic nature of precarious work, where workers labour alone, outside of a traditional workplace, creates an additional wedge between them and potential union membership. As well as the pressure of navigating a 'flexible' labour market, where the nature of work is changing, workers must also traverse the unending economic uncertainty that comes part and parcel with the business-interests-driven political mismanagement we have come to expect across the UK. Britain remains a rather hostile environment for trade unions, but moves to strengthen them should be viewed not as radical or as a regression into the bitter class politics of the twentieth century, but as an essential and, indeed, unavoidable step in creating a fairer, more prosperous and ultimately more socially cohesive society.

So how do we increase union membership? One proposal in recent years is that of automatic trade union enrolment which, while certainly strengthening unions in terms of membership, and no different in principle from being enrolled in a pension scheme, will be seen by many as either coercive or as a blank cheque to flabbier union bureaucracies. In the current context, the radical prospect of auto-enrolment would invite hostility from the usual suspects and the notion of default membership of a union with party-political affiliation would only throw petrol on that flame. Still, defaulting workers to union membership where collective agreements already exist and where there is no party-political affiliation may still be viable – with the condition that employees are free to opt-out discreetly if they so wish, in accordance with freedom of association rights. Non-affiliated auto-enrolment would lower membership costs and promote a culture where trade union membership becomes normalised. With respect to workplaces there is no recognised union, and therefore no allowance or arrangement of facility time, branch meetings, and other elements of branch-level workplace democracy, every employer must give its staff paid, unsupervised time to use for workplace democracy. For instance, if a business takes on a whole new cohort of young, precarious workers – maybe they have been recruited with the help of the Job Guarantee scheme – then that business or body should give workers this time for workplace democracy. This will be a different kind of facility time. It might be facilitated by workers, to get staff together talking. It might be used for training and discussion, led by an experienced facilitator. It might be used to address collective issues, grievances, or improvements they want, facilitated by an organiser from the union they would be members of. These meetings will be management-free, unless bosses are invited. It would be in the power of the workers.

Further steps must be taken to democratise workplaces. Board-level representation of workers – where employees elect or appoint representatives to the strategic decision-making body of companies and within the public sector – may be one way forward. The success of any enterprise, whether a private business or a public service, hinges on a collaboration between workers and executives but, unlike other European nations, the UK (like the US) follows a corporate governance system of shareholder primacy and has no means by which employees can participate in the wider decision-making processes of their employers. This drives a proximity gap between employees and more senior stakeholders, and in this model, workers are viewed remotely, as separate from decision-making which might affect them. They must accept the will of executives, appointed by shareholders, who have no sense of a worker's plight beyond their productive capacity. Bringing workers into the boardroom, in the form of elected representatives, closes this gap.

Research published in February 2021 on the Centre for Economic Policy Research's online Vox portal found that in shared governance models, such as we see in Germany and France (where worker representation slides on a scale depending on the number of workers employed by a company), 'Workers moving into a firm with worker representation experience a 4 per cent increase in wages, compared with their former co-workers moving between firms without representation.'

We need to get serious about rebalancing industrial relations by strengthening worker representation. This idea that worker representation is optional, and a matter of individual choice, while things like driving licences and birth certificates and passports and taxes are mandatory, strikes me as deeply odd, given how central labour rights have been to greater equality. Why does the government send us reminders throughout the year about registering to vote and renewing our road tax, but is happy to take such a passive role in promoting trade union membership? Why are citizens acting in a free market as consumers always deemed to be behaving rationally but when they organise as workers for better pay, eyebrows are raised? Why are multinational corporations permitted to expand indefinitely while unions are constrained by governments which are supposed to represent ordinary people's interests? And why do we accept that the Conservative Party can change the law to limit how much money trade unions can donate to the Labour Party while itself facing no such constraint where its countless faceless billionaire donors are concerned?

Why does the wisdom of the executive possess such a premium and the rights of the worker elicit such a cringe? The exploitation of workers around the world remains a serious problem and in the UK, the roll-back of labour rights and the rise of social inequality are related. If we know that the presence of trade unions in workplaces and greater representation at board level lead to better security for workers, and that this in turn leads to wider equality for their families and communities, then why are leading politicians still viewing industrial relations from the skewed perspective of the boardroom? When our leaders are so craven in the face of power and wealth that the only thing they're prepared to cross is a picket line, perhaps it's time they re-assessed their politics – or moved into another profession.

*

As you might imagine, I could write a whole book positively brimming with the fantastical changes I would attempt in a heartbeat, had I the political power or influence to do so. From offering young non-violent offenders building apprenticeships in exchange for early release and putting them to work on a radical social house building programme that makes more efficient use of their excess energy, to the legalisation of cannabis (a drug which has caused no known fatalities and is no more dangerous to mental health or wellbeing than alcohol, gambling or pornography), the tax revenue from which would fund a rights-based system of treatment for people with addiction problems, there are solutions to many of our problems if we dare to look beyond the tedious limitations of current convention. Imagine a Welfare State partly integrated with the health service in areas of deprivation, where the long-term unemployed were sanctioned to visit a counsellor or addiction specialist rather than terrorised and hounded onto the streets and an early death. Imagine a system of Universal Basic Services, which addressed the cost-of-living crisis not by depositing a basic income in our accounts every few weeks (which without economic reform would only accumulate as greater wealth at the top), but instead by adopting a notion of wellbeing which does not centre on money or earnings, but around addressing basic unmet needs such as public transport and childcare. As well as free primary and secondary education, further and higher education could be opened to all, the cost of which would be covered by a new wealth tax.

To many of you, these will seem like lofty proposals. But what if that's what we need right now? In the mid-nineteenth century, many of the distinguished moderates of their day sneered at the idea of trains ferrying people around in subterranean London; today they rightly hold the tube up as a world-class example of what Britain is capable of when it really wants to get something done. As with much of the UK's social progress, there remains a formidable impediment to transitioning from our current predicament as a deeply unequal society, still stubbornly wedded to nineteenth-century norms and traditions, to a more forward-looking, equitable and harmonious country. Which bring us to the final area – Westminster.

While I will always work for Scottish Independence – it offers greater proximity between me and the politicians I vote for – I recognise that at time of writing, that dream looks a little distant. I don't believe Britain is a bad country, nor do I think its history is any more violent or racist than any other Western nation. History is complex and we can cherry-pick whatever suits us, yet despite my sincere belief that the Union between England, Scotland, Wales and Northern Ireland is on its last legs, I also recognise the collective achievements of the home nations and the contributions we have made to the complex weft of human endeavour. It's just a shame that the great potential of Britain to be a truly world-leading beacon of social equality has been squandered by those who appear to believe this country is their own private little piggy-bank.

The final reforms I propose would target Britain's two main democratic institutions – the House of Commons and the House of Lords – where accountability must be strengthened and proximity gaps must be closed. More influential standards authorities regulating the conduct and entitlements of elected representatives, and a more proportional electoral system which reflects the will of the people are nothing short of essential for Parliament to remain relevant in the twenty-first century. As a Scot who has never voted Conservative in my life yet has lived under Conservative rule for most of it, I sympathise with all British citizens who no longer see the point in voting – the British electoral system is outdated and serves only the duopoly.

Which is why my first proposal targets citizens themselves, who feel there is little reason to participate and have taken to shrugging at talk of politics in self-defeating resignation – you are part of the problem. To reinforce a collective commitment to democracy, and better representation, automatic electoral enrolment and compulsory voting should be introduced. The arguments against it (that it would be totalitarian to force people to participate in elections) are weak, when you consider the great many tasks we are compelled to perform which are nowhere near as important as voting. If the government can compel us to attend school, enshrine in law the legal age at which we are allowed to consume alcohol, smoke cigarettes or to gamble away all of our savings, then it can also place in the statute book an obligation that we must make the effort to cast our vote every few years. People will still be free to spoil their ballots as means of protest. The jingoism around Britain's struggles for democracy does not match the level of political participation and this suits many in power, who drape themselves in the national flag and talk up the UK's world-leading democratic credentials while benefiting politically from low voter turnout and orchestrating ever more devious means of limiting how marginalised voters can participate.

It is clear whose interests are served by consistently low levels of political participation and it's not working-class people and the poor. The long-term corrosive effects on public trust in democratic institutions are evidenced in the vertical freefall of standards in Westminster. But the resulting low turnouts produce results which skew towards the interests of the affluent and wealthy and lead to vast sections of the British population dropping out of politics entirely, fomenting apathy and distrust. Full democratic participation of all social classes would create healthier political incentives and a truer picture of Britain's socio-political diversity.

Now we turn to the electoral system itself – First Past the Post. FPTP – a 'winner takes all' system which completely discounts the votes of those who didn't back the victorious candidate – is rarely representative and often leads to a government which a majority of the country did not vote for. This antiquated system is a big part of the reason why the Conservatives regard themselves as the natural party of government – under it they have dominated throughout recent history, with only brief intermissions from other parties. Previously, the main argument made in defence of FPTP was that it produced a decisive result which made governing easier, but after years of calamitous Brexit negotiations and a disastrous political response to the pandemic, as well as independence movements in Scotland, Wales and Northern Ireland, there is little evidence to support this claim. Like all the other systems we have explored in this book, the UK's electoral system exists primarily to serve the interests of a privileged minority of voters, politicians and special interests, and creates the false impression the UK is a deeply conservative country when, often, the majority which votes against the government desires a more progressive society. First Past the Post has to go. While more proportionally representative systems have their flaws (believe me, I'm Scottish), they do produce healthier incentives for parties to work together, as well as a broader sense within the electorate that there is some correlation between how we vote and who is elected. At the end of the day, how can we place proximity at the heart of democracy when election results don't even reflect the will of the people? It's the central absurdity of the British political system.

To bolster the proximity gains arising from a more proportional electoral system, we must also end the farce of hereditary privilege in the House of Lords. Hereditary peerages must be abolished and the House of Lords replaced by a new second chamber comprised of civic society, community leaders, scientists, professionals, campaigners and people with lived experience. What do hereditary or anointed peers bring to the legislative process that ordinary citizens cannot? A second chamber configured in such a manner would allow policy to be reviewed and informed by people with specific expertise as well as those who will potentially be affected. Such a reform would also engender a sense of democratic renewal, symbolic of a breakaway from the norms of the past.

To provide sufficient incentives which guard against the purchase of political power by special interests, or the misuse of political influence by those who have since left public office, a number of steps must be taken to strengthen standards in Parliament. We must put an end to secrecy over political donations; how political parties are funded has to be a matter of full transparency. This must include the indefinite exclusion of wealthy non-domiciles from contributing financially to political parties in Britain. Substantial increases in fines and other penalties for donors who break election laws may remove the moral hazards which arise when wealthy interests calculate that breaking the rules is worth the risk.

As recommended by Open Democracy, Parliament must 'accept the recommendations on dark money and transparency set out by the Committee on Standards in Public Life' and obstruct the revolving door between politics and business which creates the perverse incentives which curdle the well of public trust in democratic institutions. All MPs should be barred from taking board positions or any other consultancy or advisory roles for a minimum for five years after leaving public office and should be disallowed from seeking public office for a similar period having vacated such a role. We must place integrity at the heart of politics and create new formal and informal constraints on individuals who have, through a lack of accountability and the increasingly poor, self-serving behaviour modelled by certain leaders in recent years, lost all contact with the notion of what public service entails – serving the public. To reverse a trend which has seen entire branches of government captured by private interests, as well as the overpopulation of the private sector by former leading politicians, an unequivocal warning shot must be sent across the bow of all who believe public office is a legitimate means by which one may enrich oneself. They should instead feel empowered, encouraged even, to pursue their aspirations in the private sector exclusively, where a keen nose for self-advancement and a selective concern for the public good are surely more welcome and suited.

And so, my friends, we have reached the end of our journey. I truly hope, even where there was disagreement, that you took something from our time together. While I do strongly believe that sometimes we need to make sacrifices and fight for our beliefs, I feel equally strongly that we must have good reasons to support them and that we refine those beliefs by engaging sincerely with others who think and feel differently. I look forward to being grilled by you all in the coming months and years as we emerge from this dreadful pandemic.

Before I bid you a fond farewell, I'd like to address each of what I estimate are the various demographics reading this book. For those in public services, who feel you are constrained by the limitations or cultures in which you currently work, whether in health, policing, criminal justice, housing or social work, I implore you to recognise, and make efforts to remind yourself whenever you forget, that the struggles in which you are engaged professionally are linked to all the other struggles. Drug deaths are the final output of health inequality; rough sleeping is the final output of housing inequality; the attainment gap is the final output of educational inequality; and poverty is the final output of extreme wealth. These are the mechanisms by which class inequality is reproduced and it is often these forces, acting together upon a vulnerable person throughout the course of their life, which culminate in their social exclusion or unnecessary premature death.

When government legislation is primarily about preserving or advancing the interests of the top 20 per cent, and policies that would truly tackle inequality at its root conflict with those interests, then drug deaths, educational attainment gaps and housing crises are inevitable. It is only by weaving together these disparate struggles and presenting the wider public with a new narrative about the structural nature of inequality, in all its horrifying detail, followed by a positive vision of a better future, thus mobilising sufficient public opinion to create the political incentive necessary, that the radical change I know you wish to see can ever become possible.

To those working-class people who agree with the ideas enlarged upon in this book, to you I say this – get organised. Join a trade union. Donate to a grassroots tenants' organisation – even if you have a mortgage. Open a food bank and tell any politician looking for a photo-op to take a running jump. Go to university and shake it up. Enter the legal profession and make your values visible by retaining your accent and defending those who nobody else will. Become a journalist who is not afraid to push back against an editor. Become an editor who is not afraid to push back against an owner. Enter politics, whether professionally or as part of a campaign or social movement. Don't take shit from the poshos whose feathers need ruffling and don't take lectures from those left-wingers who confuse the isolation they experience as proof of their steadfast integrity and unshakable radicalism, rather than evidence they are ideologically fixated and stubbornly poor collaborators. Do what you feel you must do and however experience, age and responsibility change you, never fully renounce that spirit of resistance. Resist, not simply to preserve your working-class interests but to save 'informed', 'cultured', 'sophisticated' middle- and upper-class people from themselves.

And to the middle and upper classes gracious enough to walk with me to the very last page, in spite of everything I have hurled at you, it is almost time to place this book down by your bedside and forget about it. I suspect, given the pointed nature of my criticisms, that you may have an expectation that I will outline some solutions for you, or tell you what you might do better as an ally – as if describing the problem in such exhaustive detail was not hard enough!

I admit that being middle class isn't all sunshine, rainbows and reed diffusers. But you must also admit that the vague sense you have that your relative prosperity may be correlated in some way to the toil of poorer people is not exactly baseless paranoia now is it? If you want to help, you could start by holding your nose occasionally and voting for something a bit more radical – rather than professing to be politically homeless. You could stop shaking your head at single mums smoking across the road from the school and focus on the exhaust fumes expelled by the 4×4 you drove one mile to drop your children off in. You could lend your hand and your voice to community groups and enterprises which are led by working-class people, and rather than seeking to dominate, humble yourself and absorb the wisdom they have to offer – as coarsely expressed as it may be. If you're a teacher, you could stand up to your colleagues who believe placing children who misbehave in social isolation as punishment represents anything but child cruelty. If you are a doctor, you could be more public about the challenges you face working in poorer communities as a result of the uneven funding of health services. If you're a copper, you could grass up some of your colleagues now and then instead of turning a blind eye. If you are a legal professional, you could educate yourself about the impact of poverty-induced trauma and neglect on children and how it often finds expression in adolescence as 'criminal behaviour'. If you run a business, you could commit to paying your staff a little more than the living wage and if that is unaffordable, you might question why the business model you have adopted only works when you pay poverty wages. If you are a journalist, you could give more voice to people living in the toughest circumstances, asking them not only about what they have experienced but also what they think needs to be done at the political level to alleviate their suffering. If you're an opinion columnist, you could try leaving some oxygen in the room for other people to speak occasionally and give it a rest with your tedious culture war. And if you are a politician, you could ask yourself some searching questions about how the vast gulf between the prospectus of your party, its strategy and tactics, and the urgent needs of working-class people, not always factored into that calculation, can be bridged – if it can't you should be more honest with yourself and the rest of us about that.

Whatever you do, before you do it, promise me this. While you toss and turn as something elemental nags at your soul, or when you catch your reflection in the windscreen of the Lexus or on the back of one of your flat-bottomed, colour-coded sauce ladles, that you will, despite the overwhelming social, financial and professional pressure to look the other way where the fundamental unfairness of British society is concerned, occasionally ponder the most serious question of all. The question you bat away because it so discomforts you. The question you know in your heart must be addressed at some point. The only question worth asking anymore. When it comes to social inequality in Britain, what if poor people aren't the problem?

Acknowledgements

When your name's on the cover, it's easy to delude yourself that the achievement of finishing a book is yours alone. The reality is that without some form of help and support almost every day for the three-and-a-half years it took me to finish this, it may never have materialised. This work is the culmination of a mammoth team effort. The friends who picked up the slack when I was otherwise unavailable. The various editors who came on board at the different points to help me whittle down the first unwieldy manuscript I exhaustedly submitted in 2021. My colleagues at Tern Television with whom I collaborated on the documentary film sequences which are expanded upon in this book. And the courageous and generously forthcoming contributors whose stories we did our best to capture, some of which are also featured in this volume.

I must first thank my partner Rebecca and my two children Daniel and Lily for putting up with my physical and emotional absences at points due to writing or filming commitments. If it's any consolation, I'm just as difficult at work as I am at home. Thanks must also go to Rebecca's parents, Linda and Edward, for all that they have done and continue to do for us – one day I will hand-pick and personally deliver the flowers! My sister Sarah, who's always stepped in at short notice to bail me out of childcare conundrums, public transport issues, and the occasional mental health breakdown – I love you and look forward to the day I get to read your first book! Jennifer – wonderful and mysterious creature – thank you for becoming our nanny during those tough lockdown months. I can say hand on heart if not for the time and headspace your informal childcare services provided us, neither Becci nor I could have made the progress we did during the pandemic.

It's also fairly obvious that your wonderful way with the kids has had a profoundly positive impact, especially wee Lily. You put in quite a shift, and I am eternally grateful to you.

Now for some brief acknowledgements less close to home but no less important. Thanks to my agent, Vivienne Clore, for your support. Thanks to Andrew Goodfellow at Ebury for your understanding, encouragement, empathy and patience. There were points I felt like throwing in the towel and almost did. I only wish I had asked for help sooner, but you live and you learn. Paul Murphy, Suzanne Connelly and Liz Marvin, your keen editorial gaze drove me positively insane at points but almost every suggestion you made was spot on and taught me something about myself and my writing. I'd also like to thank my old publisher Gavin MacDougall at Luath Press for not only releasing me from my contract in 2018 so that I could explore this new opportunity with Penguin Random House, but also for giving me his full blessing – I owe you so much for taking a chance on me back in the day. To Harry, Ceara, Emma, Cara, Adam and the whole team at Tern, our work together helped me remain close to the action, proximity from which this book has benefited immeasurably. Thank you for allowing me to draw from our experiences on-the-ground, and for being so supportive when things were difficult. I'm not sure if there are such things as 'friends' in this industry, but you guys have certainly come closest to meeting the criteria. I'm grateful for the work you do and for the sensitivity, respect and care you bring to our brave, often vulnerable, contributors. Thanks also to BBC Scotland for your foresight in commissioning our work, and for trusting us to put on screen what we all knew in our hearts was as close to the truth as television can get.

I would like to pay tribute to all the journalists, researchers, academics, broadcasters and commentators whose work I have drawn from as second-hand research in this book. Whether statistical, reportage, testimonial or investigative in nature, the sheer depth and scope of information freely available and easy to locate (and read) was immensely helpful. I couldn't do what you guys do and therefore my work absolutely depends on yours – thank you for painstaking work, attention to detail and most of all for your integrity. When people wrongly refer to me as a journalist or an intellectual, I always feel obligated to correct them out of deep respect for your craft and expertise. I have endeavoured to acknowledge every instance where I have drawn from the work of other writers, however, the process of writing this book was chaotic throughout and I am not academically trained. If anyone should find any element of their work in this book which is not attributed to them due to oversight on my part – there are about seven different versions of this book and twice as many drafts as well as multiple drafts of many chapters – please do not hesitate to contact me directly to resolve any concern you may have.

I would also like to thank the various organisations with which I have consulted or worked with over the years, who have deepened my understanding of various issues covered here. Better Than Zero, Scottish Women's Aid and Scotland's Violence Reduction Unit in particular, as well as the various youth groups and community organisations I've encountered over the years, offer a depth of knowledge in their various fields which has been of immense help, particularly when touching subjects which must be treated with sensitivity and care.

Thanks must also go to my followers and supporters online and in the real world. It's nice to know some people actually like and appreciate you. I tend to tune out the positive comments and focus on the negative remarks, but one day maybe I will learn to internalise the many kind and compassionate things you have said or written to me over the years. My Twitter followers are like unpaid researchers, helping me source information quickly, or challenging my assumptions in ways that push me harder, so thanks for being there and for engaging with me – I really do value our interactions and appreciate your time. Hopefully I'll see you in the real world soon.

Now with that said, I would like to take some time to acknowledge who I consider to be my greatest creative collaborator and the person, besides me obviously, most responsible for this book taking the shape it has – Stephen Bennett, director of *Darren McGarvey's Scotland* and *Class Wars*. Our work together began in the winter of 2018 after an exciting but intense and exhausting year. I had never worked on a documentary film before except as a contributor and despite perhaps giving the outward appearance of confidence, I was very nervous. While I'd forgive you for assuming the jitters were caused merely by vanity – one does become slightly preoccupied with oneself from time to time – my real fear was that the subjects of our film, who had kindly agreed to share their stories with us, might come to regret placing their trust in us.

My main motivation for agreeing to co-write and present the first series was to put right something I felt films about poverty always got wrong: the story is never the chaos of the lives of the misfortunate or downtrodden – it's the systems failing them which we must always come back to. Stephen, you shared this vision from day one. You gave me my place, and I will never forget that. It did wonders for my intermittent self-esteem to be treated as a professional whose insights and opinions were valued, rather than a subject from which narrative is simply extracted for entertainment. Our work together really demonstrated for me the power of people from different social classes coming together around common goals, working on an equal footing, where the give and take, while not always easy, becomes part of the creative process rather than an impediment to it.

While I am no veteran of film-making like yourself, I've had many experiences of dealing with people in media and wish I could say they were all as positive as our work together. Your work ethic is unlike anything I have ever seen. Your diligence equally so. In you I see an artist and I think that's the level we really connected on. Whether setting up the shot, briefing me in the car on those dark mornings (thank you for all the coffees) or operating two cameras and multiple microphones simultaneously while navigating tricky terrain, you were always the safest pair of hands wherever our travels took us. From the glens of Angus to the water-logged cricket fields of Inverness, Friday night in Possilpark to the Edinburgh financial district at the height of rush hour, I was never short of reasons (or excuses) to lose my cool or what remained of my concentration. It was your presence, thoughtful direction and trust which calmed me enough to do my job to the best of my ability.

Your commitment to the contributors (I hate that word but you know what I mean) was genuine and consistent. Yes, I may be good at creating that rapport on and off camera, but yourself and the team at Tern put in many hours identifying, contacting and speaking with them, often at their most vulnerable. If they trusted me when the cameras rolled, it's because their minds were put at ease long before I arrived on set. As you know, my biggest fear is that they might, after

ACKNOWLEDGEMENTS

having taken part, see the films and feel we didn't do them justice. That we sensationalised or exploited them. I'm very proud and relieved to say that this has not happened yet and that is in large part thanks to what you and the rest of the team do with that footage in the cutting-room. I would never have been able to complete this book without our films to draw from. Lockdown (and that wee spell in rehab) simply thwarted most of my plans to get around the UK. You and everyone at Tern have gone over and above what was required of you with respect to additional support as I laboured many difficult months during the pandemic to complete this work. Whether sourcing transcripts of interviews, pointing me to sources for research or even just words of encouragement and feedback, I never felt truly alone in the isolation and because of you I managed to finish this bloody book. I hope it wasn't as difficult to read as it was to write.

A Note on Sources and
Further Reading

I drew from a wealth of articles, reports, books, shows and journals while researching this book and am very grateful to the journalists, researchers, academics, broadcasters and commentators whose work helped me to write it. Most of the interviews featured were conducted when I was working on my television series' *Darren McGarvey's Class Wars* and *Darren McGarvey's Scotland*. Articles and reports that I have cited are freely available online.

In the course of writing this book, I found the following resources incredibly informative.

BOOKS AND ARTICLES:

Blair, Tony, *A Journey* (Hutchinson, 2010)

Blyth, Mark, *Austerity: The History of a Dangerous Idea* (Oxford University Press, 2015)

Brown, Gordon, *Beyond the Crash*: *Overcoming the First Crisis of Globalisation* (Simon & Schuster, 2010)

Cameron, David, *For the Record* (William Collins, 2019)

Carnegie, Andrew, 'The Gospel of Wealth', *North American Review* (as 'Wealth'), Vol. CXLVIII (1889), available online: https://www.carnegie.org/about/our-history/gospelofwealth/

Coffin, Peter, *Custom Reality and You* (Independently published, 2018)

Dunt, Ian, *How To Be a Liberal: The Story of Liberalism and the Fight for its Life* (Canbury Press, 2020)

Eatwell, Roger and Goodwin, Matthew, *National Populism: The Revolt Against Liberal Democracy* (Pelican, 2018)

Fitzpatrick, Claire and Williams, Patrick, 'The neglected needs of care leavers in the criminal justice system: Practitioners' perspectives and the persistence of problem (corporate) parenting', *Criminology & Criminal Justice*, Vol. 17 issue: 2 (2017): 175-191

Freire, Paulo, *Pedagogy of the Oppressed* (Penguin Modern Classics edition, Penguin, 2017)

Giridharadas, Anand, *Winners Take All: The Elite Charade of Changing the World* (Penguin, 2018)

Goodall, Lewis, *Left for Dead?: The Strange Death and Rebirth of the Labour Party* (William Collins, 2018)

Hedges, Chris, *America: The Farewell Tour* (Simon & Schuster, 2018)

Kogan, David, *Protest and Power: The Battle for the Labour Party* (Bloomsbury, 2019)

Livingston, Eve, *Make Bosses Pay: Why We Need Unions* (Pluto Press, 2021)

Mattinson, Deborah, *Beyond the Red Wall: Why Labour Lost, How the Conservatives Won and What Will Happen Next?* (Biteback, 2020)

McEnaney James, *Class Rules: The Truth About Scottish Schools* (Luath Press, 2021)

Nagle, Angela, *Kill All Normies: Online culture wars from 4chan and Tumblr to Trump and the alt right* (Zero Books, 2017)

Orwell, George, 'Why I Write' (Penguin Great Ideas series, Penguin, 2004)

Pinker, Susan, *The Village Effect: Why Face-to-Face Contact Matters* (Atlantic Books, 2015)

Pluckrose, Helen and Lindsay, James, *Cynical Theories: How Activist Scholarship Made Everything about Race, Gender, and Identity – And Why This Harms Everybody* (Swift Press, 2020)

Reay, Diane, *Miseducation: Inequality, Education and the Working Classes* (Policy Press, 2017)

Richards, Steve, *The Prime Ministers: Reflections on Leadership from Wilson to May* (Atlantic Books, 2019)

Scruton, Roger, *How to be a Conservative* (Bloomsbury Continuum, 2014)

Sowell, Thomas, *Black Rednecks & White Liberals: Hope, Mercy, Justice and Autonomy in the American Health Care System* (Encounter Books, 2006)

Stewart, Mo, *Cash Not Care: the planned demolition of the UK welfare state* (New Generation Publishing, 2016)

Thatcher, Margaret, *The Downing Street Years* (HarperCollins, 1993)

Wightman, Andy, *The Poor Had No Lawyers: Who Owns Scotland (And How They Got It)* (Birlinn, 2015)

VIDEOS:

Big Think, Nicholas Christakis: The Sociological Science Behind Social Networks and Social Influence, 20 October 2019, https://www.youtube.com/watch?v=wadBvDPeE4E

YaleCourses, Lecture 1: Introduction to Power and Politics in Today's World (Ian Shapiro), 17 September 2019, https://www.youtube.com/watch?v=BDqvzFY72mg

Index

About the Author

Darren McGarvey grew up in Pollok, Glasgow. He is a writer, hip-hop artist, broadcaster and campaigner. His bestselling and acclaimed first book *Poverty Safari* was awarded the Orwell Prize for political writing in 2018.

@lokiscottishrap